Advances in Geographical and Environmental Sciences

Series editor

R. B. Singh

AIMS AND SCOPE

Advances in Geographical and Environmental Sciences synthesizes series diagnostigation and prognostication of earth environment, incorporating challenging interactive areas within ecological envelope of geosphere, biosphere, hydrosphere, atmosphere and cryosphere. It deals with land use land cover change (LUCC), urbanization, energy flux, land-ocean fluxes, climate, food security, ecohydrology, biodiversity, natural hazards and disasters, human health and their mutual interaction and feedback mechanism in order to contribute towards sustainable future. The geosciences methods range from traditional field techniques and conventional data collection, use of remote sensing and geographical information system, computer aided technique to advance geostatistical and dynamic modeling.

The series integrate past, present and future of geospheric attributes incorporating biophysical and human dimensions in spatio-temporal perspectives. The geosciences, encompassing land-ocean-atmosphere interaction is considered as a vital component in the context of environmental issues, especially in observation and prediction of air and water pollution, global warming and urban heat islands. It is important to communicate the advances in geosciences to increase resilience of society through capacity building for mitigating the impact of natural hazards and disasters. Sustainability of human society depends strongly on the earth environment, and thus the development of geosciences is critical for a better understanding of our living environment, and its sustainable development.

Geoscience also has the responsibility to not confine itself to addressing current problems but it is also developing a framework to address future issues. In order to build a 'Future Earth Model' for understanding and predicting the functioning of the whole climatic system, collaboration of experts in the traditional earth disciplines as well as in ecology, information technology, instrumentation and complex system is essential, through initiatives from human geoscientists. Thus human geoscience is emerging as key policy science for contributing towards sustainability/survivality science together with future earth initiative.

Advances in Geographical and Environmental Sciences series publishes books that contain novel approaches in tackling issues of human geoscience in its broadest sense – books in the series should focus on true progress in a particular area or region. The series includes monographs and edited volumes without any limitations in the page numbers.

More information about this series at http://www.springer.com/series/13113

Asheem Srivastav

The Science and Impact of Climate Change

Springer

Asheem Srivastav
Indian Forest Service
Gandhinagar, Gujarat, India

ISSN 2198-3542　　　　　　　　ISSN 2198-3550　(electronic)
Advances in Geographical and Environmental Sciences
ISBN 978-981-13-0808-6　　　ISBN 978-981-13-0809-3　(eBook)
https://doi.org/10.1007/978-981-13-0809-3

Library of Congress Control Number: 2018960186

© Springer Nature Singapore Pte Ltd. 2019
This work is subject to copyright. All rights are reserved by the Publisher, whether the whole or part of the material is concerned, specifically the rights of translation, reprinting, reuse of illustrations, recitation, broadcasting, reproduction on microfilms or in any other physical way, and transmission or information storage and retrieval, electronic adaptation, computer software, or by similar or dissimilar methodology now known or hereafter developed.
The use of general descriptive names, registered names, trademarks, service marks, etc. in this publication does not imply, even in the absence of a specific statement, that such names are exempt from the relevant protective laws and regulations and therefore free for general use.
The publisher, the authors and the editors are safe to assume that the advice and information in this book are believed to be true and accurate at the date of publication. Neither the publisher nor the authors or the editors give a warranty, express or implied, with respect to the material contained herein or for any errors or omissions that may have been made. The publisher remains neutral with regard to jurisdictional claims in published maps and institutional affiliations.

This Springer imprint is published by the registered company Springer Nature Singapore Pte Ltd.
The registered company address is: 152 Beach Road, #21-01/04 Gateway East, Singapore 189721, Singapore

Contents

1 **Earth in Reverse Gear** 1
   1.1 This Is What the Brundtland Commission Had Said in 1987 2
   1.2 The Observed Changes 9
      1.2.1 Chronic Droughts 9
      1.2.2 Receding Glaciers 11
      1.2.3 Acidification of Oceans 11
      1.2.4 Habitats Destroyed, and Species Driven Out
           of This Planet 12
   References 18

2 **A Glimpse of Natural Climatic History** 21
   2.1 Climate of the Past 21
   2.2 Understanding the Current Phase 26
      2.2.1 Solar Irradiance and Earth's Climate 30
      2.2.2 Radiative Forcing 32
      2.2.3 The Terra Incognita of Climate Change 35
   References 37

3 **Understanding the Warming Process** 39
   3.1 The Greenhouse Gases and Their Effects 39
   3.2 The Carbon Factor 43
      3.2.1 Data set on Carbon and Carbon Dioxide 47
      3.2.2 Properties of Carbon Dioxide 48
      3.2.3 Carbon in the Terrestrial System 54
      3.2.4 The Soil Carbon 60
      3.2.5 Man-made Carbon Source 63
      3.2.6 Carbon Dioxide in the Ocean 65
      3.2.7 Biological Calcification 68

	3.3	The Janus Faced—Nitrogen	70
		3.3.1 Nitrogen Fixation and Uptake	72
		3.3.2 Nitrogen Mineralization	72
		3.3.3 Nitrification	72
		3.3.4 Denitrification	73
		3.3.5 Major Anthropogenic Sources of Inorganic Nitrogen in Aquatic Ecosystems	74
	References		76
4	**Natures' Reaction to Anthropogenic Activities**		**79**
	4.1	Increased Vulnerability to Natural/Man-Made Disasters	79
	4.2	Population and Extreme Weather Nexus	84
	4.3	Human Species Is Under Serious Threat	93
		4.3.1 Water Distresses	95
		4.3.2 Risk to Marine Ecosystem and Corals	97
		4.3.3 Changing Temperatures and Forests	99
		4.3.4 Rising Seas	100
		4.3.5 Droughts and Fire	103
	References		107
5	**Reducing Carbon Growth**		**111**
	5.1	Introduction	111
	5.2	Can We Postpone 'Energy Apocalypse'?	114
	5.3	The Future	118
		5.3.1 The Solar Energy	118
		5.3.2 The Wind Energy	119
		5.3.3 The Hydro-/Geothermal Energy	122
		5.3.4 Energy for Batteries (Lithium: The New Gasoline)	124
		5.3.5 The Biomass Energy	125
		5.3.6 The Biogas Energy	128
		5.3.7 The Hydrogen	128
		5.3.8 Energy from Waste	129
	5.4	On Way Out	129
		5.4.1 Coal	129
		5.4.2 The Hydrocarbons	130
	5.5	The Debate on Nuclear Energy	131
		5.5.1 Environmental Effects of Nuclear Power Plants and Explosion	133
		5.5.2 Is Nuclear Energy Futuristic?	135
	5.6	Innovations	136
		5.6.1 Paradigm Shift Is Inevitable	140
		5.6.2 Bottlenecks to Sustainability	142
		5.6.3 Back to Basics	144
	References		145

Chapter 1
Earth in Reverse Gear

> *The environment is everything that isn't me.*
> Albert Einstein.

Abstract The 1972 Stockholm conference on the Human Environment acknowledged the negative effects of rapid industrialization post World War II. Subsequently, the effect of mining and burning of fossil fuels, manufacture and release of toxic chemicals and pollutants into air and water, destruction of forests, soil erosion, etc., was unambiguously brought on record by the Brundtland Commission in 1987 with a clear warning for humanity to either change its development strategy or be prepared to face the consequences in near future. Unfortunately, the situation has further worsened due to unchecked population growth, consumption pattern, use of fossil fuels and release of greenhouse gases, deforestation and pollution. There are incontrovertible evidences that temperatures today are more than 0.8 °C above pre-industrial levels and the sea level has been rising @3 mm per year. The world is currently witnessing frequent and severe natural disasters including droughts, floods, cyclones, fires and mudslides. Scientists have warned that in the event of temperature rising to 5 °C above pre-industrial levels, 50% of animal and plant species will become extinct, 30% of coastal wetlands would be inundated, terrestrial ecosystems will shift from 'carbon sink' to being 'carbon source', many small island states will suffer from storm surges, and nearly three billion people will be under severe water stress.

Keywords Ozone layer · Fossil fuels · Natural disaster · Acid rain · IPCC
Climate change · Sea level rise · Threatened ecosystems · Receding glaciers
Threatened species · Greenhouse gases · Drought and climate change
Population growth and climate change · Brundtland Commission

1.1 This Is What the Brundtland Commission Had Said in 1987

- Over the past century, the use of fossil fuels has grown nearly **30-fold**, and industrial production has increased more than **50-fold**. The bulk of this increase, about three-quarters in the case of fossil fuels and a little over four-fifths in the case of industrial production, has taken place since 1950. The annual increase in industrial production today is perhaps as large as the total production in Europe around the end of the 1930s. Into every year, we now squeeze the decades of industrial growth and environmental disruption that formed the basis of the pre-war European economy.
- The impact of growth and rising income levels can be seen in the distribution of world consumption of a variety of resource-intensive produce. The more affluent industrialized countries use most of the world's **metals** and **fossil fuels**. Even in the case of food products, a sharp difference exists, particularly in the products that are more resource-intensive.
- The "greenhouse effect", one such threat to life-support systems, springs directly from increased resource use. The burning of **fossil fuels** and the cutting and burning of forests release carbon dioxide (CO_2). The accumulation in the atmosphere of CO_2 and certain other gases traps solar radiation near the Earth's surface, causing global warming. This could cause sea level rises over the next **45 years** large enough to inundate many low-lying coastal cities and river deltas. It could also drastically upset national and international agricultural production and trade systems.
- Another threat arises from the depletion of the atmospheric **ozone layer** by gases released during the production of foam and the use of refrigerants and aerosols. A substantial loss of such ozone could have catastrophic effects on human and livestock health and on some life forms at the base of the marine food chain. The 1986 discovery of a hole in the ozone layer above the Antarctic suggests the possibility of a more rapid depletion than previously suspected.
- A variety of air pollutants are killing trees and lakes and damaging buildings and cultural treasures, close to and sometimes thousands of miles from points of emission. The **acidification** of the environment threatens large areas of Europe and North America. Central Europe is currently receiving more than one gram of sulphur on every square metre of ground each year. The loss of forests could bring in its wake disastrous erosion, siltation, floods and local climatic change. Air pollution damage is also becoming evident in some newly industrialized countries.
- In many cases, the practices used at present to dispose of toxic wastes, such as those from the chemical industries, involve unacceptable risks. **Radioactive wastes** from the nuclear industry remain hazardous for centuries. Many who bear these risks do not benefit in any way from the activities that produce the wastes'.
- More than 11 million hectares of **tropical forests are destroyed per year**, and this, over 30 years, would amount to an area about the size of India. Apart from the direct and often dramatic impacts within the immediate area, nearby regions are affected by the spreading of sands or by changes in water regimes and increased risks of soil erosion and siltation.
- Chemicals represent about 10% of total world trade in terms of value; some 70,000–80,000 chemicals are now on the market—and hence in the environment. The figure is only an informed estimate because no complete inventory has been done. Some 1000–2000 new chemicals enter the commercial market each year, many without adequate prior testing or evaluation of effects. The negative environmental impacts of industrial activity were initially perceived as localized problems of air, water and land pollution. Industrial expansion following the World War II took place without much awareness of the environment and brought with it a rapid rise in pollution, symbolized by the Los Angeles smog; the

1.1 This Is What the Brundtland Commission Had Said in 1987

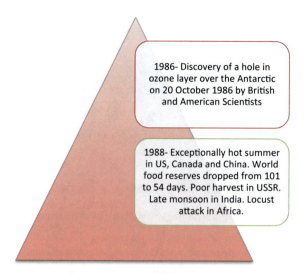

Fig. 1.1 Watersheds in understanding climate change (*Data Source* Ceres 1990 Vol 1, 25 Sept–Oct)

proclaimed 'death' of Lake Erie; the progressive pollution of major rivers like the Meuse, Kibe and Rhine; and chemical poisoning by mercury in Minamata. These problems have also been found in many parts of the Third World as industrial growth, urbanization and the use of automobiles spread. Industrialized countries still suffer from 'traditional' forms of air and land pollution. Levels of sulphur and nitrogen oxides, suspended particulates and hydrocarbons remain high and in some cases have increased. Air pollution in parts of many Third-World cities has risen to levels worse than anything witnessed in the industrial countries during the 1960s (Brundtland Commission 1987).

It has now been established beyond an iota of doubt by the collective works of thousands of scientists across the globe that the Earth is warming. Two important events have helped in establishing this fact (Fig. 1.1). First was the discovery and confirmation of a hole in the ozone layer above Antarctica by a team of American and British Physicists in 1986 (Monier 1990). The second was summer of 1988 with exceptionally hot and dry weather affecting the food production in the USA, Canada, China, Russia, India and other countries and regions including locust attack in Africa. The food reserves dropped from 101 to 54 days. To avoid scepticism, the United Nations thought it prudent to reconsider the statistical data and compare with the past climate records and prepare forecasting models. The Intergovernmental Panel on Climate Change (IPCC) was asked to prepare a report to this effect. The IPCC submitted its first report in 1990 inter alia with following assertions:

1. Emissions resulting from human activities are substantially increasing the atmospheric concentrations of the greenhouse gases: CO_2, methane, CFCs and nitrous oxide. These increases will enhance the greenhouse effect, resulting on average in an additional warming of the Earth's surface. CO_2 has been responsible for over half the enhanced greenhouse effect; long-lived gases would require imme-

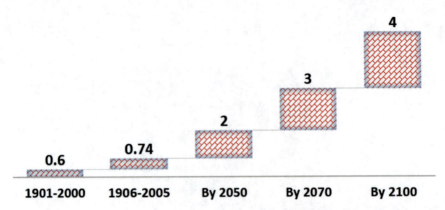

Fig. 1.2 Global average temperature (degree Celsius) changes so far and anticipated. *Source* IPCC 2007

diate reductions in emissions from human activities of over 60% to stabilize their concentrations at today's levels.
2. If we continue living, consuming and producing as we do now, the global temperature increase will be 0.3 °C per decade (ranging from 0.2 to 0.5 °C) from the coming century. This is greater than that seen over the past 10,000 years; warming will be greater on land areas than over the ocean and more pronounced in the northern hemisphere.
3. The rising levels of ocean have been observed in the Maldives and the Guyana coast. A total of 14 island states accounting for 700,000 inhabitants in the Pacific and the Indian Oceans are under the threat of submergence due to sea level rise.
4. There were many uncertainties in the predictions particularly with regard to the timing, magnitude and regional patterns of climate change, due to our incomplete understanding of: sources and sinks of GHGs; clouds; oceans; polar ice sheets.

More than four decades after the Stockholm conference (in 1972), the global environmental scene has worsened as more than three billion people have been added and the cumulative impact of increasing human population, changing consumption pattern, over consumption by some and under consumption by others and the avoidable wastage has begun to produce adverse consequences. Humans currently dominate more than half of biological production through agriculture, forestry, industries and other activities (Fig. 1.3). In doing so, they have increased the release of GHGs on the one hand and have reduced the ability of ecosystems to absorb these gases that are generated as a by-product or end product. As a result, the current global temperatures are higher than they have ever been during the past one thousand years and there are incontrovertible evidences that temperatures today are more than 0.8 °C above pre-industrial levels (Fig. 1.2). The fact that global average temperature continues to increase can be gauged from the recent data which estimates that the average global temperature for 1969–1971 was 13.99 °C that rose to 14.43 °C during 1996–1998, a gain of 0.44 °C (Brown 2000).

1.1 This Is What the Brundtland Commission Had Said in 1987

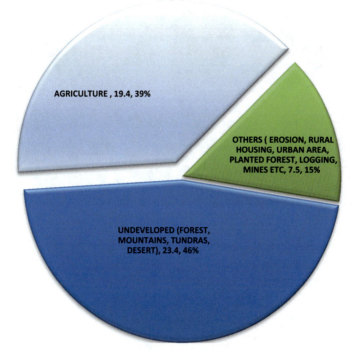

Fig. 1.3 Use of ice-free area of the world (million square miles) (*Data Source* Foley 2014)

The main reason for the current rise in temperature is attributed to high levels of carbon dioxide concentration as well as to some other greenhouse gases such as nitrous oxide, methane and halocarbons that continue to trap infrared radiations in excess vis--vis pre-industrial time and the resulting warming causes long-term climate change. Under normal circumstances, these GHGs act like a blanket to keep the Earth's surface 20 °C warmer for the survival and growth of living entities (Lacis 2010). Unfortunately, the projections, based on current warming trends, indicate that the human actions have already committed the world to a 2 °C warming and any temperature increase beyond this may have disastrous consequences (Schneider 1989). The concentration of carbon dioxide has now been ascertained more accurately for the past 650,000 years from the analysis of Antarctic ice cores. During that time, CO_2 concentration varied between 180 ppm (glacial cold) and 300 ppm (interglacial warm) (Sigman and Boyle 2000). In contrast, the CO_2 concentration shot up by more than 80 ppm in the twentieth century, which would have otherwise taken 5000 years under natural circumstances.

The impacts of this warming are unevenly distributed across the planet. For example, temperature rise is greater at the poles with some regions of the Arctic warming 0.5 °C in last 30 years (World Bank 2010). Snow and ice melting in the Arctic and Greenland have become common; cold days and nights have become less frequent

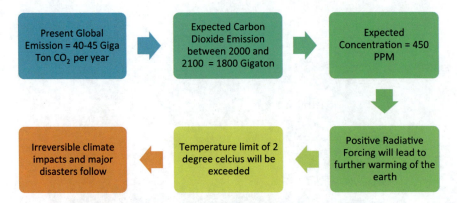

Fig. 1.4 Illustration of CC impacts if carbon dioxide concentration exceeds the limit

while intensity and frequency of heatwaves have increased. Cyclones, droughts and floods occur more frequently in different parts of the world.

The scientific community, while categorizing climate change impacts in five categories, viz.

i. Threatened ecosystems and rare/unique species;
ii. Extreme weather events;
iii. Extent of impacts;
iv. The economic impact;
v. Significant range of discontinuities.

Strongly believes that the future impacts of 2 °C rise (above pre-industrial level) in global temperature will be severe in terms of coastal erosion, water availability (to more than billion people in mid-latitudes and semi-arid low latitudes, mainly in Asia and Africa) (Yoon 2013) and extinction risk (to nearly 25% plant and animal species). In the worst-case scenario of temperature rising to 5 °C above pre-industrial levels, 50% of animal and plant species will become extinct, 30% of coastal wetlands would be inundated, terrestrial ecosystems may shift from 'carbon sink' to being 'carbon source', many small island states will suffer from storm surges, and nearly three billion people will be under severe water stress (Fig. 1.4). Evidences reported in IPCC 2007 indicate that the thawing of Greenland and Antarctic ice sheets as well as that of permafrost and mountain glaciers is faster than expected. It also suggests that the droughts in West Africa and drying of Amazon rainforest may be more likely than thought earlier. While future scenarios are based on past and present data analysis and modelling experiments, in reality the predicted changes in different regions may occur in jumps and shifts. As scientists understand the ambiguities in climate change forces and impacts with better clarity, sudden, intense and unpredictable events continue to occur in different parts of the world.

The instrumental data of sea level change during twentieth century shows an increase of 1.7 mm per year. Similarly, satellite altimetry data gathered since 1993

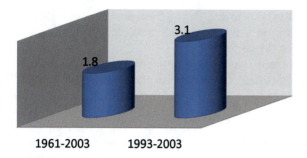

Fig. 1.5 Average sea level rise (in mm) (*Source* IPCC 2007)

indicates that sea level has been rising @3 mm per year which is significantly higher than the average during the previous half century. However, the rise is not uniform around the world and there are sharp variations (IPCC 2007). While in some regions rate of rise is several times higher than average, there are other regions where sea levels are falling. This is mostly due to inconsistent changes in temperature and salinity and the resultant changes in ocean circulation. It is estimated that on average, between 1961 and 2003 thermal expansion contributed about 25% of the observed sea level rise (Fig. 1.5), while melting of snow accounted for nearly 50% (Draper and Kundell 2007). However, during the period between 1993 and 2003, the rise in temperature and melting of ice accounted for nearly 50% sea level rise (IPCC 2007).

Global sea level is predicted to rise further and as per the IPCC report on emission scenario, by the mid-2090s, the global sea level is projected to reach 0.22–0.44 mm above 1990 levels. Scientists have also pointed out that the recent rate of global sea level rise has significantly departed from the average rate of two to three thousand years and is rising more rapidly at one-tenth of an inch per year increasing the vulnerability of millions of people living close to the coastal areas across the globe (Glick 2004). Rising sea levels produce a cascading effect, and it has been estimated that every inch of sea level rise may result in 96 inches or eight feet of horizontal submergence of shoreline with salt water intruding into freshwater aquifers affecting the agriculture and drinking water potential.

With a sharp and sudden population growth from about 1 billion in 1800 to over 7 billion in 2011, the humanity has transited from agriculture-based economy to industrial economy propelled by fossil fuels. The prophecy of Thomas Malthus in 1798 that human population grows faster than the food supplied until war, famine and disease reduce the number has gone wrong. And the credit for this goes to the discovery of fossil fuel, chemical fertilizers, high yielding food plants, pesticides, disease control, improved health and sanitation among other things. Experts also believe that the use of fossil fuel and population explosion are inextricably linked and the present level of human population is attributed to the discovery and exploitation of fossil fuels (Meyerson 2003). While there is no data to prove the foregoing statement, the converse is true. And that is—that the unprecedented of growth of human population has surged per capita fossil fuel consumption from less than 0.01

metric ton to 1.1 metric ton resulting in global emissions of carbon dioxide that grew from 8 million metric tons to 6518 million metric tons between 1800 and 1999, an 820-fold increase (Meyerson 2003).

Besides, the ever-rising footprints of agriculture has caused tremendous loss of several natural ecosystems including the prairies of North America, rainforests of Brazil, boreal forest and tropical forests. Hunger of land for agriculture and development has led to diversion of precious and irreplaceable natural forests. On a global scale, nearly 60 per cent of recent deforestation is attributable to agriculture sector (*farming is among the largest contributors to global warming emitting more GHGs than transport sector*), 20% to logging (including mining and petroleum) and 20% to household sector for fuel-wood, all GHG-enhancing activities (Population, Environment and Development 2001). Farming sector has also sucked up underground water supply, dried up several perennial rivers through damming and canal for irrigation, caused salinity ingress in coastal areas and polluted water bodies with run-off fertilizer, insecticides and pesticides.

The double whammy of increasing population and prosperity is driving the world crazy for high energy diet that mainly comes from egg, chicken, fish, pigs, lamb and beef. At present, only 55% of the world's crop is being fed directly to the people, about 36% is fed to the livestock, and the rest 9% goes into biofuels and other industrial products (Foley 2014). How the world will feed 9 billion mouths in 2045 is a dilemma for which there is no solution at present especially considering the fact there is a trend towards consumption of animal protein and therefore food production will have to far outpace population growth particularly in developing world where per capita protein demand will increase by 103.6% by 2050. We do not realize the fact that for every kilo calorie of grain that is fed to the animals, only 40% is converted into of milk or 22% is converted into eggs or 12% into chicken or 10% into pork or only 3% into beef. In other words, the loss of energy from grain to meat conversion is extremely high, and scientists and policymakers will have to work on new ideas to ensure minimization of wastage (Fig. 1.6) and efficient use of grain for undernourished people of the world.

The overall impact[1] of climate change is, nevertheless, different in developed world though the direction of impact varies. A rapidly growing rich global consumer society of 1.7 billion is currently responsible for the over use of biological products, fossil fuels, minerals and metals and other diverse products and services than ever before, while 2.8 billion poor struggle to survive on less than USD 2 per day. Affluent nations like USA, Canada, Britain, France, Germany, Japan and Italy represent only a tenth of global population but consume over 40% of Earth's fossil fuels as well as most of the world's commodities (Fig. 1.7) and forest products. With every dollar increase in per capita GDP, there is 1.4 kg increase in per capita CO_2 emissions (Olivier et al. 2014). In other words, higher socio-economic status has equal or more detrimental impact on the environment. The ecological footprint of an individual in a high-income country is at least six times higher than in low-income countries.

---

[1]Impact $= f$ {Population $\times$ Affluence(or Poverty) $\times$ consumption pattern $\times$ Technology}.

1.2 The Observed Changes

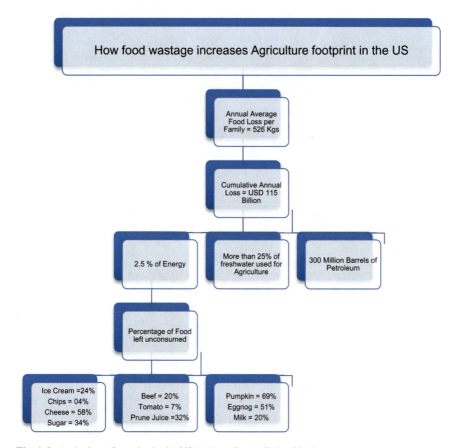

Fig. 1.6 Agriculture footprint in the USA (*Data Source* Foley 2014)

1.2 The Observed Changes

1.2.1 Chronic Droughts

Drought, whether meteorological, hydrological, agricultural or environmental, has been a regular feature considered as a temporary aberration from normal climatic conditions. However, the current spates of frequent, severe and widespread droughts with devastating consequences have been reported in many regions. Persistent, higher than usual temperatures have attributed to the exacerbated retention of high moisture in the atmosphere, and high rainfall in certain areas, causing flash floods, erosion and mudslides. As a consequence, certain other areas suffer from less than normal rainfall or no rains at all. Also, the areas suffering from flash flood and soil erosion face drought as less moisture is retained. Recent researches on the impact of atmosphere and ocean in altering the global weather and climate pattern indicate that sea surface

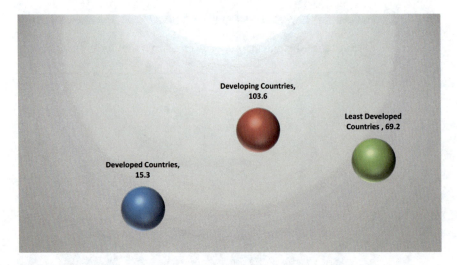

Fig. 1.7 Per cent increase in daily protein demand by 2050 (*Data Source* Foley 2014)

temperatures in the central and eastern tropical Pacific Ocean are substantially higher than usual during El Nino event and substantially lower during La Nina. These temperature fluctuations are strongly linked to major climate fluctuations around the globe, and their impact lasts for more than 12 months (Rojas et al. 2014). For example, the strong El Nino event of 1997–1998 was followed by prolonged La Nina phase that lasted from mid-1998 to early 2001. Climatic variations apart, there are many countries that have, over the years, overdrawn groundwater as well as surface water (river damming, etc.) for agriculture and industrial development. Such countries suffer most in the event of prolonged absence of or shortage of monsoon. One of the conspicuous examples is Australia that has suffered from millennium drought beginning 1995 and lasting until 2009. Catalonia region of Spain (*about 70% of Spanish water is used for agriculture primarily due to poor irrigation system and water thirsty crop*) was so badly affected by drought in 2008 that water was imported from France by ship.

Populated and monsoon-dependent countries like India suffer from drought almost every year. Large-scale withdrawal of water for irrigation and human consumption in Haryana, Punjab, Rajasthan and Delhi between 2002 and 2008 has depleted groundwater to the extent of 108 cubic kilometres. Syria follows India in terms of groundwater loss. Between 2006 and 2011, Syria suffered worst drought and severe shortage of water in Tigris and Euphrates river basins.

1.2.2 Receding Glaciers

By definition, a glacier is a large,[2] long-lasting mass of ice that is formed on land by several years of ice accretion. Of the two main types of glaciers, the **alpine** glaciers are found in mountain terrain and the **continental** glaciers cover large areas of continents and are associated with ice ages. Glacier records are extremely useful in understanding climatic variability due to rapid changes in their mass and thus contribution to the rapid sea level rise. As humans continue to impact the Earth through the burning of fossil fuels and other activities, predicting future requires understanding of past climate variability. Glaciers around the world, including those in the Himalaya, are a unique laboratory for helping unlock the secrets of our complex climate system. Similarly, the Glacial National Park of the USA, encompassing 4000 km^2, had nearly 150 glaciers in the mid-nineteenth century. However, only 25 remain now facing further challenges from global warming impacts that threaten to engulf all of them by 2025 lest the climatic situation improves (Key et al. 2002). Data of the snout positions of a few of the thousands glaciers in Himalayan region highlights the worse that they have been in a general state of decline over at least the past 150 years. It also suggests that the rate of retreat has been increasing in the past decade and the melting glaciers are filling Himalayan mountain lakes and river systems too quickly, threatening millions of lives with unpredictable and colossal floods and landslides. If the glacial retreat continues over the long term (several decades), as is expected in a greenhouse-gas-warmed climate, the amount of water melted will decrease and the flow of rivers in southern Asia will become less reliable and eventually diminish causing potential widespread water shortages leading to disastrous consequences.

1.2.3 Acidification of Oceans

There are three principal ways in which oceans affect the climate:
i. Transfer of huge amount of water vapour between sea and air;
ii. Transport of considerable amount of heat from tropics to the poles;
iii. Being huge reservoir of CO_2 (50 times more than atmosphere), they maintain Earth's heat balance.

Oceans capacity to hold CO_2 depends on following factors:
i. Chemical properties of CO_2 in sea water;
ii. Presence of 'Biological Pump'[3] that transports CO_2 from the sea surface to deep ocean;

---

[2] Glaciers are the largest reservoir of fresh water on earth and second largest reservoir of total water (after ocean).
[3] Biological Pump—Phyto-planktons on ocean surface that carry huge quantity of CO_2 to the ocean floor where it stays for hundreds of years.

iii. Rate and pattern of ocean circulation.

The oceans absorb 22 million tons of CO_2 per day, thus proving valuable service to the survival of mankind (Feely et al. 2006). There is no other mechanism to absorb CO_2 at this rate, and this invaluable service of oceans has a high ecological cost. Since the beginning of the Industrial Revolution, the ocean has become 30% more acidic and the speed of this change is unparalleled in known human history. Carbonic acid formed by dissolution of CO_2 in water decreases the availability of carbonate making it difficult for marine organisms including corals, mussels, snails, and sea urchins to construct their hard parts out of calcium carbonate minerals. Decreased calcification in marine organisms makes them vulnerable to extrinsic factors such as erosion and pollution. In a few invertebrates and fish, CO_2 accumulation and lowered pH may result in acidosis, a build-up of carbonic acid in body fluids, leading to lowered immune response, metabolic depression and asphyxiation. Ocean acidification may also affect marine food webs and lead to significant changes in commercial fish stocks, threatening food security for millions of people thereby adversely impacting a multi-billion dollar industry. Similarly, coral reefs generate billions of dollars annually in tourism, which may be at risk as reef area diminishes and corals become more prone to diseases. In addition, reefs will provide shorelines that are more vulnerable to erosion and flooding.

1.2.4 Habitats Destroyed, and Species Driven Out of This Planet

Each of the biological species on this planet is a storehouse of matchless substances for the sustenance of the planet. Of the fourteen biomes, the two most species rich biomes are the tropical forests and the coral reefs. The former contain at least half of the world's species but are under serious threats (Fig. 1.8) largely from conversion to other land uses for human use, while the latter are experiencing increasing levels of pollution and over exploitation. A whole range of plant-derived dietary supplements, phyto-chemicals and pro-vitamins that assist in maintaining good health and combating disease are now being prescribed as food supplements by experts and pharmaceutical companies. In addition to human use, many plants and plant products are also used for treatment of captive animals in zoos and circuses. Because of the major contributions that these plants make in terms of health support, financial income, cultural identity and livelihood security, these plants have become the cornerstone of both human and veterinary medical systems worldwide. It is estimated that 484 animal and 654 plant species have already been driven to extinction during the past four centuries and almost 1400 tropical plants and 500 marine organisms which yield chemicals for cure of cancer will soon be driven to extinction before their potential can be assessed or tapped (Srivastav and Srivastav 2015). Unconfirmed statistics indicate that worldwide between 35,000 and 70,000 plant species are used as medicines and food supplements of which 9000 are threatened (McNeely and Mainka 2006). Europe alone shares around 25% of world trade in medicinal plants, and the demand is growing.

1.2 The Observed Changes

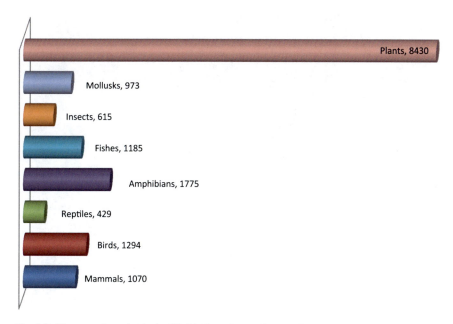

Fig. 1.8 Threatened species in the World (*Data Source* Earth Policy Institute 2008)

Commercial and non-commercial trade in wild animals and their parts and products for food, therapeutic and other uses not only affects animal populations and their habitat but also the local and indigenous communities who rely on locally obtained medicines for their own basic needs. For example, population of Hose's langur endemic to Borneo in Indonesia has sharply declined because of its hunting for bezoar stones used in traditional medicines. Similarly, populations of Saiga antelope for horn; seals for genitalia; pangolin for scales; bear for salts derived from gallbladders, musk deer for musk; rhinoceros for horn and tiger for bone are on decline. Every component of tiger's body including hair, skin, meat, testes, tail, stomach, nose, whiskers and bones holds some therapeutic value in treating a litany of ailments and nearly twenty million freshwater turtles and tortoises are consumed annually by consumers in China, Lao PDR, Cambodia, Vietnam, Malaysia, Thailand, India, Nepal, Bangladesh and Sri Lanka.

With the present rate of human population growth and consumption, the coming decades are expected to witness biological holocaust. Environmental degradation and growing pollution are already threatening biodiversity and ecosystem stability, and many Asian and African nations are currently at a stage when their physical and biological systems may not be able to meet even their basic needs (potable water, fodder, energy, shelter and food) of the growing population. Increasing frequency of landslides, floods, droughts, soil loss and other human incited natural disasters continues to compound the miseries of poor communities' particularly marginal and poor farmers as well as the landless labourers (Tables 1.1 and 1.2).

Table 1.1 Projected impacts of mean annual temperature changes (degree Celsius)

	0.8–1.8	1.8–2.8	2.8–3.8	3.8–4.8	4.4–5.8
Africa			Semi-Arid/Arid area increase by 5–8%		
	75–250 million people face water stress	350–600 million people face water stress	More people face water stress		
		10–15% sub-Saharan species at risk of extinction	25–40% sub-Saharan species at risk of extinction		
Asia	2–5% decrease in wheat production in India	5–12% decrease in rice production in China			
		Up to 2 million people at risk of coastal flooding each year	Up to 7 million people at risk of coastal flooding each year		
	0.1–1.2 billion people face water stress	Additional 0.2–1 billion people face water crisis			
Australia/New Zealand	Bleaching of Great Barrier Reef				
	3000–5000 more heat-related deaths per year				
	Decreasing water security				
Europe	+5 to +15% water availability in North	+10 to +20% water availability in North			
	0 to −25% water availability in South	−5 to −35% water availability in North			

(continued)

1.2 The Observed Changes

Table 1.1 (continued)

	0.8–1.8	1.8–2.8	2.8–3.8	3.8–4.8	4.4–5.8
Latin America			Threat of extinction to 25% central Brazilian savannah tree species		Threat of extinction to 45% Amazonian tree species
		Many tropical glaciers disappear	Many mid-latitude glaciers disappear		
		10–80 million people face water stress	80–180 million people face water stress		More people face water stress
North America	Decrease in space heating and increase in space cooling				
		Between 5 and 20% increase in crop yield			70–120% increase in forest fire in Canada
					3–8 times increase in heatwave days in some cities
Polar region			20–35% reduction in permafrost area		10–50% Arctic tundra replaced by forests
					15–25% polar desert replaced by tundra
					Between 20 and 35% decrease in average Arctic sea ice area

Data Source Adopted from Dessler (2002)

Naturally and human-induced factors have contributed significantly to the degradation of coastal and marine ecosystems, most of which are irreversible. The biggest damage to these ecosystems has been caused by silt brought down by soil erosion due to the deforestation and poorly planned developmental activities, encroachment, excess sedimentation due to poor land use practices, industrial effluents, urban sewage, over fishing, clear felling in coastal forests and mining of coral rocks for building material, and extraction of coral sands for cement production. Clearance of land for farming and other uses has also resulted in sedimentation and siltation, impacting especially on mangrove and reef areas. Many industries including petrochemical, cement, shrimp farming, tanneries, slaughterhouses and other chemical processes contribute solid waste and wastewaters to the environment, often without adequate treatment.

A large number of coastal cities and towns have become vulnerable to hazards and perturbations that have become frequent and more intense. For example, cyclonic

Table 1.2 Anticipated impact by the end of twenty-first century

Africa	Two important changes may happen in African region. One, there will be an increase in arid and semi-arid lands and two, low-lying coastal areas will be adversely affected thereby increasing the cost of adaptation
Asia	Climate change impacts due to rapid urbanization, industrialization and economic development are expected to enhance the pressures on natural resources. Densely populated coastal regions in South, East and South-east Asian region will be prone to flooding and erosion due to sea level rise. Besides, the freshwater availability is expected to decline in major river basin areas. Large-scale disturbances in hydrological cycle may increase human deaths and diseases
Australia and New Zealand	Population growth and unabated coastal development coupled with sea level rise will increase the risk of coastal flooding as well as severity of storms. Agriculture and forest production will be hard hit in the eastern New Zealand and SE Australia due to increased risk of drought and fire. Coastal and marine biodiversity in the Great Barrier Reef will undergo considerable decline
Europe	Most of the European nations will be severely impacted by climate change events especially the coastal and mountainous areas. Coastal countries will be subjected to increased and frequent flooding and erosion, whereas mountain regions will face decline in snow cover, glacier retreats and extensive loss of biodiversity. Countries in southern region will be worst affected due to reduced water availability, hydropower generation and agriculture production
Latin America	While most of the tropical countries in Latin America are at a very high risk of species loss, higher than normal temperature in many parts will increase soil water loss consequently leading to replacement of tropical forests by savannah ecosystem in eastern Amazonia
North America	The likelihood of cities, towns and villages facing heatwaves with increased intensity and duration is expected to rise with serious consequences on human health. Mountain regions in the western part will see more flooding in winters and reduced water flow in summer, thereby increasing the risk of flash floods and drought
Polar regions	The impact will be more pronounced in polar regions. For instance, the reduction in thickness and extent of glaciers and ice sheets will profoundly change natural ecosystems with detrimental effects on many organisms including migratory birds and mammals
Small islands	Communities living in small island nations will be the first to bear the brunt of sea level rise. Many small island nations will face severe drinking water stress due to changes in rainfall pattern. A large number of islands are expected to face exacerbated inundation, storm surge, erosion and other coastal hazards that will threaten infrastructure, human settlements and livelihood support system

winds and storm surges cause inundation of low-lying areas of coastal region, drowning human beings and livestock, eroding beaches and embankments and destruction of vegetation, reducing soil fertility and polluting drinking water, extensive damage to infrastructure, dwellings, communication systems. Heavy and prolonged rains due to cyclones repeatedly cause river floods and submergence of low-line areas leading to loss of life and property.

> **Box 1.1 How Acid Rain impact us**
>
> Increasing consumption of fossil fuel for industrial and household activities, automobiles and other anthropogenic reasons increases the acidity in the rainwater, a phenomenon called 'acid rain'. Charles Crowther and Arthur Ruston recorded, for the first time in 1911, a pH of 3.2 in the industrial town of Leeds, UK. Subsequently, low pH value was recorded at many places (Murray 1919). In November 1964, storm water in NE USA had a pH of 2.1 (Waananen et al. 1971); in April 1974, the downpour in the west coast of Norway and Iceland had a pH values of 2.7 and 3.5, respectively.
>
> Under normal concentrations and pressures of CO_2 in the atmosphere, the rainfall has a pH of 5.65 which is primarily due to the formation of carbonic acid (H_2CO_3), a weak acid. Increase in atmospheric acidity is caused by sulphur oxides and oxides of nitrogen injected into the atmosphere by industries, automobiles and other anthropogenic factors. There are two types of mechanisms that bring down these pollutants:
> - By wet deposition through the process of acid rain; and
> - Dry deposition in the form of dust and particulate matter.
>
> **Impact of acid rain and acidification on freshwater ecosystem**
>
> - Fish population is severely affected below pH 5.6, with decline in reproductive capacity, egg hatchability and egg survivability;
> - Respiratory and metabolic disturbances in molluscs, insects, crustaceans, fish and amphibians;
> - Reduced growth rate in fish and amphibians;
> - Growth of benthic mosses is accelerated;
> - Benthic invertebrates, zooplankton and phytoplankton try to become more acid tolerant;
> - Decline in species diversity;
> - Rooted aquatic plants are eliminated, and rate of decomposition is reduced.
>
> **Impact of acid rain on terrestrial ecosystem**
>
> - Leaching of nutrients in humus layer of soil is accelerated thereby reducing productivity;
> - Nitrification, nitrogen fixation and root nodule numbers are drastically reduced;

- Symbiotic balances between plant and mycorrhizal fungus are disturbed;
- Acceleration of cuticle erosion making leaf tissues susceptible to pathogens and saprophytes;
- Foliar loss of nitrogen, calcium, magnesium and phosphorus increases.

References

Brown, Lester Russell. 2000. *State of the world: A Worldwatch Institute report on progress toward a sustainable society*. New York: WW Norton.
Brundtland Commission. "World commission on environment and development". 1987. A threatened future. In *Our common future*, 5–39.
Ceres 1990. The FAO Review. Vol. 125 Sept–Oct. Forecast Famine (Page 15) by Francoise Monier.
Dessler, Andrew, and Lowman Student Center Theater. 2002. The science of climate change.
Draper, Stephen E., and James E. Kundell. 2007. Impact of climate change on transboundary water sharing. *Journal of Water Resources Planning and Management* 133 (5): 405–415.
Earth Policy Institute. 2008. *Threatened species in major groups of organisms, 2007*. Rutgers University.
Feely, Richard A., Christopher L. Sabine, and Victoria J. Fabry. 2006. Carbon dioxide and our ocean legacy. *Marine Life* 1: 2.
Foley, Jonathan. 2014. A five step plan to feed the world. *National Geographic* 225 (5): 27–59.
Glick, Daniel. 2004. The big thaw. *National Geographic* 18.
IPCC. 2007. *Climate change 2007: The physical science basis. Contribution of Working Group I to the Fourth Assessment Report of the Intergovernmental Panel on Climate Change*, ed. Solomon, S., D. Qin, M. Manning, Z. Chen, M. Marquis, K.B. Averyt, M. Tignor and H.L. Miller. Cambridge University Press, Cambridge, United Kingdom and New York, NY, USA.
Key, Carl H., D. B. Fagre, and R. K. Menicke. 2002. Glacier retreat in Glacier National Park, Montana. *Satellite Image Atlas of Glaciers of the World, US Geol. Surv. Professional Paper* J365–J375.
Lacis, A. 2010. CO_2: *The thermostat that controls earth's temperature*. NASA: Science Brief.
McNeely, Jeffrey A., and Sue Mainka. 2006. The future of medicinal biodiversity. *IUCN* 165.
Meyerson, Frederick A.B. 2003. Population, biodiversity and changing climate. *Advances in Applied Biodiversity Science* 4: 83–90.
Murray, J. 1919. British Association for the Advancement of Science. Meeting. Report of the British Association for the Advancement of Science.
Olivier et.al. 2014. *Trends in global CO_2 emissions: 2014 Report*. JRC Technical Note number: JRC93171.
Population, Environment and Development. 2001. The Concise Report: United Nations Publications, Room DC2 0853 New York NY 10017 USA, 68.
Rojas, Oscar, Yanyun Li, and Renato Cumani. *Understanding the Drought Impact of El Niño on the Global Agricultural Areas: An Assessment Using FAO's Agricultural Stress Index (ASI)*. 2014.
Schneider, Stephen H. 1989. The greenhouse effect: Science and policy. *Science* 243 (4892): 771–781.
Sigman, Daniel M., and Edward A. Boyle. 2000. Glacial/interglacial variations in atmospheric carbon dioxide. *Nature* 407 (6806): 859–869.
Srivastav, Asheem, and Suvira Srivastav. 2015. *Ecological Meltdown: impact of unchecked human growth on the earth's natural systems*. The Energy and Resources Institute (TERI).

References

Waananen, Arvi O., David Dell Harris, and Robert Charles Williams. 1971. *Floods of December 1964 and January 1965 in the Far Western States; Part 1 Description*. No. 1866-A. US Govt. Print. Off.

World Bank. 2010. *World Development Report 2010: Development and Climate Change*. International Bank for Reconstruction and Development/The World Bank.

Yoon, Chin Saik. 2013 Future imperatives of practice: The challenges of climate change. In *Sustainable development and green communication*, 58–77. Basingstoke: Palgrave Macmillan.

Chapter 2
A Glimpse of Natural Climatic History

Those who cannot remember the past are condemned to repeat it.

Abstract The phenomenon of climatic regime on this Earth is a dynamic process that has evolved over millions of years under the influence of its own internal dynamics as well as changes in the external factors. There is strong scientific evidence that Earth has witnessed cycles of ice ages and warmer periods during its evolutionary history and these climatic changes have been responsible for speciation and extinction of many plants and animals on this planet. Analysis of ice core has provided evidence of warming and cooling within long ice ages. But the current climate change phenomenon is attributed to human influence since 1750 and is concerned with radiation imbalance due to factors such as long-lived greenhouse gases, ozone, water vapour, surface albedo (land use changes and snow cover changes) and aerosols. Indiscriminate use of coal and petroleum derivates after the onset of Industrial Revolution and especially the upsurge after World War II has brought in significant and possibly irreversible changes in Earth's energy balance.

Keywords Environment · Radiation balance · Albedo
Palaeocene–Eocene Thermal Maximum (PETM) · Radiative forcing
Milankovitch cycle · Ice core · Foraminifera · Interglacial

2.1 Climate of the Past

The word 'environment' signifies the flow or exchange of energy between living beings as well as between living and non-living entities under a given set of conditions. Flow or exchange of energy, on the other hand, is a dynamic process which started much before the appearance of Homo sapiens on this planet; e.g. events such as plate tectonics, ice ages, change in sea levels, alterations in the global climate pattern and the origin and evolution of species have continuously changed the energy dynamics of the world. Our knowledge and understanding of such changes are limited due to scanty and elusive evidences. Nevertheless, what is well known is the fact that such events occur due to a combination of several factors. For example, it is believed that during Palaeocene epoch, i.e. around 56 million years ago, the Earth was much warmer than today and the Atlantic Ocean had not fully opened up (Kunzig and Block 2011). This facilitated the movement of man, animals and plants from Asia to North America through Europe and Greenland (Kunzig and Block 2011). The culmination of Palaeocene witnessed pulling apart of Europe and Greenland and opening up of North Atlantic Ocean. During this period, known as Early Eocene, there was massive release of carbon dioxide possibly of volcanic origin which resulted in further warming of the Earth by 6 °C extending from tropics to higher latitudes and deep oceans (National Research Council 2011).

One of the most plausible hypotheses for this carbon spike is that most of it came from large deposits of methane hydrate[1] (methane hydrate *consists of a water molecule forming a casing around a single molecule of methane*). Since methane hydrates are stable only in a narrow range of cold temperatures and high pressures, an initial warming during Palaeocene–Eocene transition (Box 2.1) or Palaeocene–Eocene Thermal Maximum (PETM) due to volcanic or some other action might have melted hydrates releasing methane that warms the Earth twenty times more than carbon dioxide. The methane so released subsequently oxidized into CO_2 and continued warming the planet for a long time. As the ocean absorbed excess CO_2 that was warming the planet, the water also became acidified. The evidence of this is found in the deep-sea sediments of arid Bighorn[2] basin where red rust bands of oxidized soil mark the sudden warming that occurred 56 million years ago. Geologists and Palaeontologists today believe that the excess carbon that was absorbed is equivalent to the Earth's reserves of oil, coal and natural gas. Analysis of carbon isotopes, extracted from Atlantic seafloor near Antarctica in 1991, indicates that huge amount of carbon had entered into the ocean and atmosphere in a short span of a few centuries during Palaeocene–Eocene Thermal Maximum (PETM) some 56 million years ago (Kunzig and Block 2011) and acidified the ocean. The acidified ocean in turn dissolved calcium carbonate shells of corals, clams and foraminiferans on a massive scale. Simultaneously, atmospheric carbon dioxide was also dissolving into rain droplets which leached calcium from rocks and washed them into sea where it combined with carbonate to form calcium carbonate which was eventually deposited as limestone at the bottom of the sea.

---

[1] Large deposits of methane hydrate are still found under the Arctic tundra and under the seafloor.
[2] Bighorn basin is a hundred mile long arid plateau east of Yellowstone National Park in Northern Wyoming, USA.

2.1 Climate of the Past

> **Box 2.1**—Palaeocene–Eocene Thermal Maximum {PETM} (Jansen et al. 2007)
>
> PETM refers to a period between Late Palaeocene and Early Eocene epochs that occurred approximately 55 million years ago. This period of unusually high temperature is believed to have lasted for less than 0.1 million years, and the environmental impact of warming was felt at all latitudes and in both the terrestrial and the aquatic systems. The ^{13}C isotopes in marine and continental recordings indicate that a sufficiently large mass of carbon with low ^{13}C concentration must have been released into the atmosphere that subsequently lowered the pH of the ocean and dissolved seafloor carbonates. It is estimated that nearly 1 to 2×10^{18} g of carbon was released during the PETM period and the probable sources for this carbon could be methane (CH_4) obtained from decomposition of clathrates on the seafloor, carbon dioxide from volcanic activity or oxidation of sediments rich in organic matter.
>
> The PETM, which modified the Earth ecosystems, has been thoroughly examined as it is believed that the current warming due to rapid release of carbon into the atmosphere by humans is also somewhat similar. Although there is still too much uncertainty in the data to derive a quantitative estimate of climate sensitivity from the PETM, the event is a striking example of massive carbon release and related extreme climatic warming.

The primary driving force for climate modification throughout the evolutionary history of the Earth has been the ***radiation balance*** of the Earth, i.e. the quantum of radiations received by the Earth from Sun and the quantum radiated back to the atmosphere and beyond. Incoming solar radiations, on the other hand, are affected by the changes in Earth's orbit around the Sun, and the amount of solar radiations received at each latitude varies with each season. In addition to influencing the incoming solar radiation, the Earth's orbital changes are believed to be responsible for regular cycles of ice ages during the past three million years. Also, there are strong indications that warmer times have also occurred during most of the 500 million years of the climate history when Earth was free from ice sheets (Jansen et al. 2007). While there is lack of consensus on the number of [3]ice ages during the past 2.5 million years or more, but it has been widely accepted that the climatic conditions of this planet waxed and waned between ice ages and interglacial (warm) period. It is believed that during Mid-Pliocene[4] (about 3.3–3.0 million years), the average global temperatures were 2–3 °C above pre-industrial temperatures for a long period.

Natural change in climatic regimes has impacted the evolutionary history of living species, and the life on Earth has witnessed at least five periods during which huge numbers of living species vanished forever, primarily due to changes in climate and sea level. The first of these five extinction periods started around 485 million

---

[3]Glick (2004).
[4]IPCC (2007), Chap. 6 (6.3.1).

years ago when the sea level rose continuously due to very high temperatures and high atmospheric carbon dioxide leading to massive destruction of marine flora and fauna. The second one occurred after 120 million years (365 million years ago) when marine species vanished due to changes in sea level and loss of oxygen. The third extinction which happened some 251 million years ago is supposed to be the largest so far when more than 96% of all species disappeared. The fourth one, known as end Triassic extinction, occurred 205 million years ago due to intense volcanic activity and leading to loss of 76% living creature mostly marine. The last known extinction (that occurred some 65 million years ago) is the most interesting because it signalled the end of 75% to 80% of all species including dinosaurs, which had dominated the land for 140 million years. Probably between 75 and 80% of all species disappeared during this time that occurred some 65 million years ago.

Scientists in the University of Sheffield, UK, believe that the fourth extinction in the Late Triassic was the result of global warming (McElwain et al. 2009). Their argument is based on the stomatal count of 18 different groups of fossil plants that existed between Triassic and Jurassic periods. It was found that the number of stomata in these plants decreased progressively and then recovered shortly. A comparison in the stomatal number of fossil and living plants of same species revealed that the amount of CO_2 during Late Triassic quadrupled over a period of a few hundred years and the temperature increased by 3–4 °C. It took more than 150,000 years for the Earth to absorb this excess carbon and allowed the planet to cool.

The Earth's climatic history is better understood from about 75,000 years ago when the first of two ice ages—[5]Wisconsin I and Wisconsin II—had set in. Wisconsin II reached its coldest about 20,000 years ago, and at that time the Laurentide was the largest ice sheet that covered northern part of North America. The other ice sheets covered Scandinavia, the British Isles and eastern Siberia in the northern hemisphere. The average temperature of the Earth's surface was 6 °C cooler than today with Central Europe and much of Asia permanently frozen. Paleo-climatologists have unearthed signs of ancient climate and climatic fluctuations by analysing a wide array of sources such as glacial ice and moraines, stalagmites from caverns, tree ring, pollen grains, corals, dust and sand dunes and the microscopic shells of organisms buried in deep seas. Glacial ice cores are like annual growth rings of a tree which provide the best indicators of atmospheric events. Fresh snow has a density of 100 kg/m^3 or less. As the snow compresses by further accumulations of snow above it, fragile snow crystals are transformed into spherical particles with density reaching up to 500 kg/m^3 after a year or more. If the cold conditions prevail for longer durations, the increasing pressure creates larger crystals with density reaching up to 830 kg/m^3, which join together and trap air in the form of a bubble. These bubbles sealed in the ice provide an insight into abrupt (regional events of large amplitude) climatic changes such as a few degree temperature variations within several decades. Like ice cores, the width of tree rings and its density is extremely useful to determine past temperature and moisture changes with great accuracy. Similarly, analysis of annual rings (like trees the density of which depends on sea surface temperature) produced by deep-sea corals provides indicators for nutrient content and mass exchange from

---

[5]Glaciers and the Environment UNEP/GEMS Environment Library No 9, Nairobi UNEP 1992.

the surface to deep waters showing abrupt variations characterized by synchronous changes in surface and deep-water properties. Fossilized shells of foraminiferans, a microscopic marine invertebrate, are exceptionally useful as they contain chemical signature of water temperature that can be dated with great accuracy. Pollen grain, due to their microscopic size and light weight, has vast spatial spread in a given time frame. Their studies indicate that tropical lowlands were on average 2–3 °C cooler than present, with strong cooling effects in Central America and northern South America and weak cooling (<2 °C) in the western Pacific Rim (Jansen et al. 2007).

Ice cores from Greenland first obtained and analysed in the 1960s hold some of the best records of temperature changes, precipitation pattern and atmospheric conditions during past ten thousand years. Greenland ice has revealed that within the long cold stretches, there was short burst of warming and cooling. These records also show that atmospheric CO_2 varied in the range of 180–300 ppm over the glacial–interglacial cycles of the last 650,000 years (Fig. 2.1). It is also believed that nearly 10,000 years ago, the global temperature and precipitation were similar to those of today and many mountain glaciers were as small as they are now. The longest ice core thoroughly analysed so far is from Vostok Station in Antarctica. The 2521-m-long ice core covering a time span of 160,000 years suggests a relationship between ice ages and astronomical rhythms (Barkov et al. 1977). Its analysis also helped scientists to establish baseline levels for CO_2 to 270–290 ppm (parts per million). Scientists have also found evidence of a sharp 32 feet rise in global sea level some 19,000 years ago due to melting of ice in northern hemisphere. During the ice ages, the North Atlantic Ocean conveyer weakened triggering a series of events that ultimately led to warmer temperatures in the southern hemisphere and colder temperatures in the north. Throughout the last ice age that lasted from about 70,000 to 11,500 years ago, mountains in Ireland were covered with glaciers and vast ice sheets blanketed North America, Europe, parts of Russia and Antarctica.

Besides, several ice ages that changed the climatic history of Earth, the world has also witnessed the Little Ice Age between early seventeenth to the mid-nineteenth century that repeatedly increased the glaciers. However, after the mid-nineteenth century, the length and mass of many glaciers have been reportedly shrinking at a much faster rate possibly due to enhanced greenhouse effect of post-industrial era (Wang and Yang 2014).

As a young student of science in the early 1970s, we were made to believe that climatic changes of significance occurred only over periods as large as geological epochs and therefore we should be more concerned about day-to-day weather. Nevertheless, there were some eminent scientists in those days who were convinced that strange things were happening to the weather around the world during the past few decades. Europe witnessed severe winters in 1963, North Africa and India faced unprecedented droughts in 1973, and there was excessive ice in polar seas in 1968 (Le et al. 2007). Derek Winstanley, a London-based meteorologist, who analysed rainfall records of Sahel region in North Africa found declining rainfall pattern since the 1960s. While comparing records of North Africa and the higher latitudes in northern hemisphere, he suggested that the downward trend in rainfall was in some way connected with atmospheric over the arctic zone. Winstanley also examined rainfall

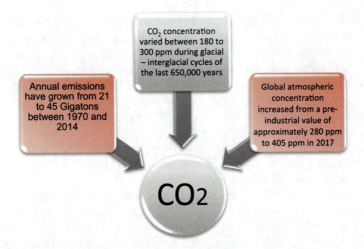

Fig. 2.1 Carbon dioxide release since Industrial Revolution has surpassed its past interglacial concentration which fluctuated between 180 and 300 ppm

records from Mauritania to north-west India comparing the average rainfall (computed over five years centred on the year 1957 and 1970) in Bikaner and Jodhpur, India, Khartoum in Sudan, Agades in Niger, Tessalit and Gao in Mali, Nouakchott and Atar in Mauritania. He concluded that average rainfall centred on 1970 in all the stations was less than half the amount recorded in 1957. In other words, regions north of the old monsoon belt showed signs of aridity possibly due to shifting of monsoon belt southwards.

2.2 Understanding the Current Phase

Events such as plate tectonics, ice ages, change of sea levels, alterations in the global climate pattern and the origin and evolution of species have continuously changed the face of the world. That the early human populations became conscious of their surroundings is evidenced from the stone artefacts which reflect high level of prudence and knowledge of non-living materials as well as other living beings. It also reflects acute human power of observation, competence in dealing with stone and sensitivity to the surroundings. The knowledge and skill of generating fire facilitated the impact of human beings on their immediate surroundings in the beginning which later extended to distant environs.

A combination of stone and fire probably helped early hominids in hunting animals, clearing forests and grasslands, fishing and using wood as well as bones and hides of animals for effectively exploiting their environment during the later phases of Stone Age. The analysis of early rock art demonstrates good understanding of plant and animal world. For example, burial of dead with flowers and red ochre (o-cur) and with ibex horn to mark the burial place has been recorded from comparatively early times in the mountains of Iran and Iraq to the north-west of Indian subcontinent. Archaeological studies including those of pollens, charred remains of wood and food grains do indicate that humans started impacting their environment during the Stone Age itself. However, the extent and quantum of this impact are difficult to assess due to spatial and temporal reasons.

> **Box 2.2—The Great London Smog of 1952—The first visible impact of Industrial Revolution**
> During November/December of 1952, heavy snowfall in and around London forced people to burn large quantities of coal to keep their houses warm. Under normal circumstances, smoke would rise into the atmosphere and disperse with wind velocity. But during that period, an anticyclone that was hanging over London created an inversion, where air close to the ground got warmer than the air higher above it. Beneath the inversion of the anticyclone, light wind stirred the saturated air upwards to form a layer of fog 100–200 m in extent. The net result was massive smog with water droplets of the fog and the smoke from innumerable chimneys.
> As more and more warm smoke came out of the chimney, it got trapped along with particles and gases emitted from factories in the London area as well as pollutants which the winds from the east had brought from industrial areas on the continent. On each day during the smog period, 1000 tons of smoke particles, 2000 tons of carbon dioxide, 140 tons of hydrochloric acid, 14 tons of fluorine compounds and 370 tons of sulphur dioxide were released. The impact of London smog was immense and led to the death of 4000 people, and large number suffered from breathing problems. In addition, many livestock perished, and people were not able to travel for several days.
> In order to ensure that such disastrous situations do not recur, several corrective measures were taken by the government. A series of laws were brought to avoid a repeat of the situation. This included the Clean Air Acts of 1956 and 1968. These acts banned emissions of black smoke, and decreed residents of urban areas and operators of factories must convert to smokeless fuels.
> *Source:* Met Office (2015) http://www.metoffice.gov.uk/learning/learn-about-the-weather/weather-phenomena/case-studies/great-smog

The earliest reference to the use of coal as fuel by humans is from the geological treatise '*On Stones*' by Theophrastus where he describes different marbles and mentions two types of coal, one which retains its heat and can be rekindled by fanning and the other that was ignited by the heat of the Sun, especially when sprinkled with water (Fortenbaugh et al. 2016). Englishmen started using coal for fuel as early as ninth century which was subsequently banned during the reign of King Edward I (1272–1307). Nevertheless, Richard II (1377–1399) revoked the ban and introduced taxation on extraction and use of coal to minimize its use. This was succeeded by

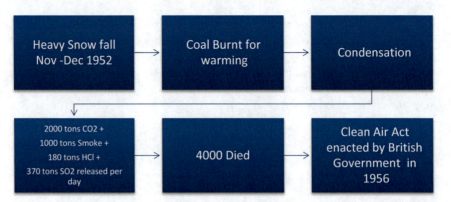

Fig. 2.2 Illustration of London Smog of 1952 and its impact

strict regulatory measures for use of coal by Henry V (1413–1422). Unfortunately, by sixteenth century, much of the natural forests in England were cleared to meet the growing requirement of fuel, timber, shipbuilding and farmland. Englishmen were left with no option but to use coal as a substitute for wood energy. With the onset of Industrial Revolution in eighteenth century, the consumption of coal shot up rapidly and the smoke started spewing out sulphur dioxide, nitrogen oxides, particulate matter and other hazardous substances to the detriment of humans and the environment. As the Industrial Revolution spread to other countries so did the problem of obnoxious gases and particulate matter. In December 1930, 60 people were killed by air pollutants in the Meuse Valley in Belgium. Investigations revealed that sulphur dioxide concentration was in the range of 25,000–100,000 microgram per cubic metre as against the current WHO standards of 40–60 microgram per cubic metre. In October 1948, 20 people died and almost 14,000 fell ill due to high concentration of sulphur dioxide (1400–5500 microgram per cubic metre) in air. The city of London was badly hit by horrifying smog in December 1952 (Box 2.2) due to high levels of particulate matter (4 billion micrograms per cubic metre). This led to the enactment of Clean Air Act of England in 1956 in the aftermath of Great Smog which fell over London city in December 1952 killing almost 4000 people and stopping trains, cars and public events (Fig. 2.2).

Petroleum (Latin petra means 'rock' and oleum means 'oil'), an important source of carbon dioxide, is a highly inflammable fossil fuel (Fig. 2.3) that occurs beneath the Earth surface and has been in human use for hundreds of years. In the early phases, people collected this yellow to black coloured oil when it bubbled to the surface while digging well for water or drilling holes for brine. In the early phase of its discovery, only kerosene was extracted from crude oil and used for lighting purposes. Its utility as commercial fuel was not realized till the middle of the nineteenth century when people started separating crude oil into gasoline, kerosene, naphtha, natural gas, lubricating oil, paraffin wax and asphalt derivatives. Subsequently with the advancement of technology, other derivates were obtained and put into commercial use mainly for

2.2 Understanding the Current Phase

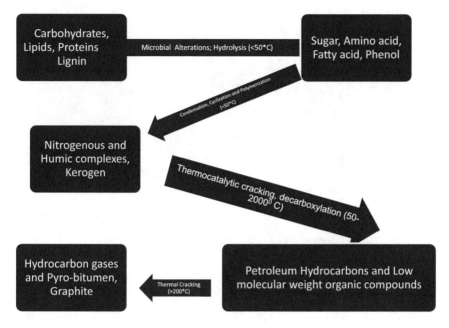

Fig. 2.3 Illustration of organic matter conversion to fossil fuel

transportation, industries and power generation. Each barrel of crude oil (42 gallons) produces around three hundred metric tons of carbon dioxide each day. Imagine the amount of carbon or carbon dioxide that has been released into the atmosphere by trillions of barrels of crude oil that has been extracted so far and will be released in future as well.

After World War II, the extraction and use of crude oil and coal increased at a rapid pace as the competition for reconstruction and economic development became top priority for most of the war-affected countries as well as countries that got independence from foreign rulers. The world was divided into industrialized, developing, centrally planned economies and capital surplus oil-exporting nations. Industrialized countries were high producer and consumer of primary energy and dominated the global energy market. The developing countries were poor in both production and consumption as they had limited resources to invest in exploration, extraction, transmission and infrastructure for oil exploration and relied mainly on coal for energy production. Centrally planned economies, on the other hand, maintained a balance between production and consumption and offset poor growth in crude oil production by increasing natural gas and coal production. Second half of the twentieth century witnessed massive increase in crude oil production and increase in price after the formation of OPEC. In a nutshell, coal and petroleum dominated the second half of twentieth century and did considerable damage to the environment by releasing carbon dioxide and other greenhouse gases (Fig. 2.4). Unfortunately, the use of coal,

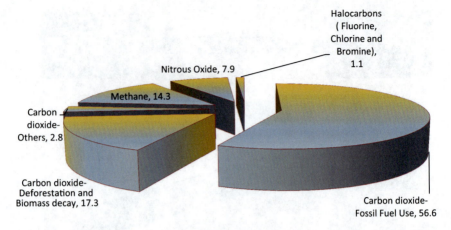

Fig. 2.4 Percent share of GHG's (*Data Source* IPCC 2007)

oil and natural gas has not stopped despite the fact that most of the nations are aware of the adverse consequences.

2.2.1 Solar Irradiance and Earth's Climate

The solar energy is radiated in the form of radiations of various wavelengths including gamma rays, X-rays, UV rays and visible light.[6] Most of the gamma and X-rays are absorbed by the thermosphere, while most of the UV rays are absorbed by the stratosphere. The energy reaching the top of Earth's atmosphere each second is of the order of 1370 watts per square metre surface area facing the Sun during daytime. This value of 1370 watts per square metre of total solar irradiance is based on thousands of measurements taken between 1902 and 1957 and subsequently by a number of scientists (Altomonte 2008). However, the average solar irradiance (Fig. 2.5) over the entire planet is 342 Watt/square metre per second (Donald 2014).

The burning energy of the Sun (solar energy) does miracle on Earth by providing energy that governs the physical and chemical processes in the atmosphere including converting the carbon dioxide from the air and water as well as from soil or sea into energy-rich carbohydrates and oxygen (Schaub and Thomas 2011). The oxygen so generated supplements the oxygen that was produced by primitive bacteria of ocean some two billion years ago, to permit living beings on the Earth to develop and evolve. In fact, bacteria (especially those inhabiting human body) are the only animals whose extinction can threaten the biological viability of man. Human beings can possibly survive extinction of all other animals except bacteria.

---

[6]Total solar energy intercepted by the earth per day $= 3.67 \times 10^{21}$ calories.

2.2 Understanding the Current Phase

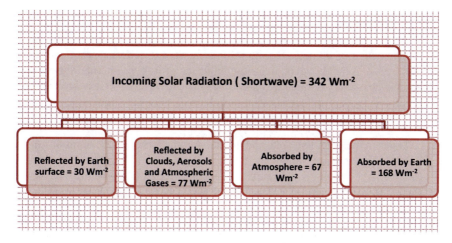

Fig. 2.5 Incoming solar radiation (short wave) = 342 Wm^{-2} (*Source Data* Herman et al. 2012; Jansen et al. 2007)

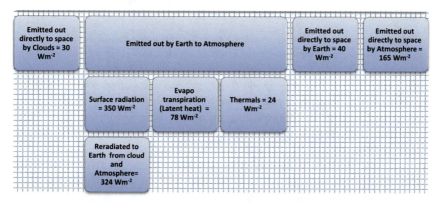

Fig. 2.6 Outgoing long-wave radiation from atmosphere, cloud and Earth = 235 Wm^{-2} (*Source Data* Herman et al. 2012; Jansen et al. 2007)

Nearly, fifty per cent of the Sun's radiation is either absorbed or radiated back by the atmosphere and clouds. Whatever reaches the ground is converted into infrared (heatwave) radiation and radiated back towards space (Fig. 2.6). Since 99% of atmosphere is either nitrogen and oxygen (neither absorbs much heat), most of the infrared radiation is absorbed by water vapour, carbon dioxide and other gases. It is only because of these heat absorbers in the atmosphere that Earth has life and most we do not freeze.

2.2.2 Radiative Forcing

The phenomenon of climatic regime on this Earth is a dynamic process that has evolved over millions of years under the influence of its own internal dynamics as well as changes in the external factors invariably called '**forcing**'. There are primarily two natural factors that influence radiation balance (i.e. the quantum of incoming and outgoing radiations):

1. Changes in the incoming solar radiation due to factors such as alterations in the Earth's orbit or in the Sun itself;
2. Change the fraction of solar radiation that is reflected (the albedo). The albedo (Fig. 2.6) can change by fluctuations in cloud cover or small particles called aerosols or land cover; and

In addition to the above natural factors, there is another factor that is primarily anthropogenic in nature and that is:

3. Alterations in the long-wave energy radiated back to space (e.g. by changes in greenhouse gas concentrations).

The current climate change phenomenon that is attributed to human influence since 1750 is concerned with radiation imbalance due to forcing or external factors such as long-lived GHG (carbon dioxide, nitrous oxide, methane and halocarbons), ozone, water vapour, surface albedo (land use changes and snow cover changes) and aerosols. The term '**radiative forcing**' indicates as to how the energy balance between the Earth and the atmosphere is influenced when factors that affect climate are altered. Radiative forcing (Fig. 2.7) is usually quantified as the '*rate of energy change per unit area of the globe as measured at the top of the atmosphere*' and is expressed in units of '**watts per square metre**'. When radiative forcing from a factor or group of factors is evaluated as **positive**, the energy of the Earth–atmosphere system will ultimately increase, leading to a warming of the system. In contrast, for a **negative** radiative forcing, the energy will ultimately decrease, leading to a cooling of the system. The combined anthropogenic radiative forcing has been assessed as +1.6 W/m^2 indicating that since the onset of Industrial Revolution the humans have exerted a warming influence on the climate.

The spherical shape, daily rotation and annual revolution of Earth make low latitude regions gain more energy than middle and high ones, and to correct this energy imbalance large-scale heat energy flows from equatorial region towards the poles through circulation patterns in the atmosphere and the oceans. The energy which goes into evaporating water from ocean and Earth is released when water vapour condenses in cloud. This energy, known as **latent heat**, is responsible for atmospheric circulation which in turn facilitates the ocean circulation. The Earth's climate (the average weather or mean and variability of temperature, wind and precipitation) primarily depends on the amount of solar radiation absorbed by the Earth and its atmosphere, and the amount that gets reflected (Fig. 2.8), which is a measure of the Earth's albedo (Maharjan and Joshi 2013). The greater the albedo, the colder is the

2.2 Understanding the Current Phase

Fig. 2.7 Radiative forcing

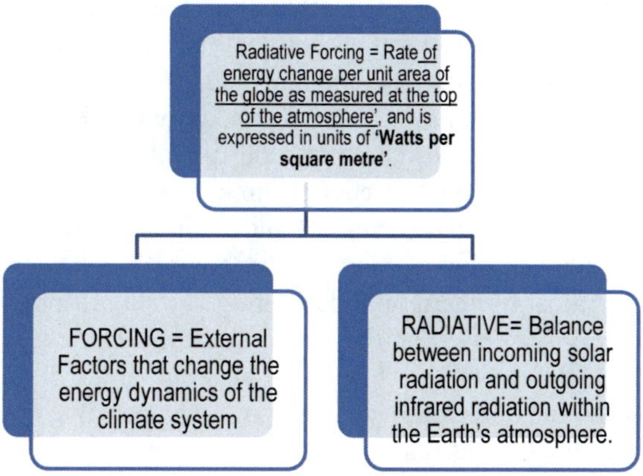

- Radiative Forcing = Rate of energy change per unit area of the globe as measured at the top of the atmosphere', and is expressed in units of '**Watts per square metre**'.
  - FORCING = External Factors that change the energy dynamics of the climate system
  - RADIATIVE = Balance between incoming solar radiation and outgoing infrared radiation within the Earth's atmosphere.

Atmospheric Zone	X- Rays	Gamma rays	UV Rays	Infra Red rays
Magnetosphere (beyond 900 Kms) - An airless protective shield created by the Earth's magnetic field.				
Exosphere (from nearly 450 to 900 Kms above earth surface) – No gases found. It is a kind of exit zone to space.				
Thermosphere (85-450 Kms above earth surface) – the upper atmosphere or Ionosphere begins here. Temperature in the zone increases with height because of electrically charged atoms due to solar radiations				
Mesosphere (50 to 85 kms above earth surface) – Temperature drops to as low as -113 *C as height in this zone increases.				
Stratosphere (15-50 kms above earth surface)- Most of the ozone is concentrated in this region absorbing UV rays from the Sun and causing increase in temperature.				
Troposphere (Upto 15 Kms above earth surface- Contains 80% of atmospheric mass including all of water vapour and dust.				

Fig. 2.8 Earth's screen

Earth. The orbit of Earth around the Sun and its orientation in space are changing regularly. Earth's orbit around the Sun expands and contracts between circular and elliptical paths in a 100,000- and 400,000-year cycle. Earth's tilted spin axis wobbles like an unsteady top, gradually making almost a full circle in space in 19,000- and 23,000-year cycle. The tilt of the spin axis increases and decreases in 41,000-year cycle. These rhythmic changes are believed to be responsible for distribution of sunlight over the Earth's surface and setting ice age. Milankovitch (Box 2.3) advanced a theory of ice ages based on the long-term variation in the solar radiation received at critical northern latitudes during certain seasons of the year (Loutre 2003). His theory was based on the premise that the ever changing shape of the Earth's orbit around the Sun, the changes in the value of the tilt of the earth's axis and the slight wobbling of the Earth on its axis were responsible for the long term variation in solar radiation received by Earth. The available evidence in the form of fossilized shells of foraminifera (microscopic marine organisms) which contain chemical signature of water temperature has helped researchers in tracing the climate back for millions of years. Because of the elliptical orbit, the tilt and the wobble, the Earth is at certain times of the year, closer to the Sun than the others.

Box 2.3—Milankovitch Theory
Milankovitch advanced the theory of ice ages based on the long-term variation in solar radiations received at critical northern latitudes during certain seasons of the year. He suggested that variation was produced as a result of ever-changing shape of Earth's orbit around the Sun by the changes in the value of the tilt of the Earth's axis and wobbling of the Earth on its axis. The Earth's orbit changes its shape from circular to elliptical and back once in 93,000 years, and the angle of tilt of the axis varies between 22.1° and 24.5° every 41,000 years. The axis also precesses as it slows completing one precession every 21,000 years.

The Milankovitch theory proposes that periodic changes in parameters of the orbit of the Earth around the Sun modify the seasonal and latitudinal distribution of incoming solar radiation (called insolation) at the top of the atmosphere. Earth's orbit around the Sun and its orientation in space change regularly. These changes facilitate the timing of ice ages by affecting the distribution of sunlight over the Earth's surface. The obliquity or tilt of the Earth axis varies between 22.05° and 24.50° with a cycle of around 41,000 years. These cycles are known as Milankovitch cycles. The changes in eccentricity of the Earth's orbit alone modulate the Sun–Earth distance and have limited impact on global and annual mean insolation. It, however, affects the **intra-annual** changes in the Sun–Earth distance and significantly modulates the seasonal and latitudinal effects. The theory states that ice ages are triggered by minima in summer insolation near 65°N, enabling winter snowfall to persist all the year and therefore accumulate to build northern hemisphere glacial ice sheets. For example, the onset of the last ice age, about 116 ± 1 ka, corresponds to a 65°N mid-June insolation about 40 W m^{-2} lower than today.

Milankovitch cycles

The Milankovitch cycles are the Earth's orbital changes around the Sun that drives the ice age cycles. These cycles change the amount of solar radiation received at each latitude in each season (but hardly affect the global annual mean), and they can be calculated with astronomical precision. There is still some discussion about how exactly this starts and ends ice ages, but many studies suggest that the amount of summer sunshine on northern continents is crucial: if it drops below a critical value, snow from the past winter does not melt away in summer and an ice sheet starts to grow as more and more snow accumulates. Climate model simulations confirm that an ice age can indeed be started in this way, while simple conceptual models have been used to successfully 'hindcast' the onset of past glaciations based on the orbital changes. The next large reduction in northern summer insolation, similar to those that started past ice ages, is due to begin in 30,000 years.
(Reference—Jansen et al. 2007)

Variations in Earth's orbit over the past millions of years are also responsible for ice ages from time to time. This has been explained in Milankovitch theory, and the ice age cycles can be predicted with great precision. Carbon dioxide concentration also plays an important role in the ice ages. Ice core data from Antarctica shows that CO_2 concentration is low during glacial times (190 ppm) and high in interglacial period (280 ppm).

2.2.3 The Terra Incognita of Climate Change

There are two distinct but little-understood events in the global climate dynamics—the El Nino and the La Nina. The El Nino (pronounced as El Ninyo) or the infant (in Spanish) was first noticed in the 1500s by fishermen of South America who began to wonder about a current of unusually warm water which they intermittently encountered near the coast during Christmas time. The Spanish speaking and curious fishermen related this strange phenomenon with the birth of the Christ and thus named the warm water as El Niño, which means 'the infant' in Spanish. The El Nino is a component of El Nino Southern Oscillation (ENSO) where the water of the Pacific Ocean and the atmosphere above interact with each other in such a way that affects the weather pattern in others parts of the world. It is a phenomenon which still remains terra incognita for the scientist community.

Under normal conditions, trade winds blow towards the west, across the Pacific, pushing *warm surface water* away from the South American coast towards Australia, the Philippines, Malay Archipelago and New Guinea (this large pool of *warm surface water* continues to enlarge in size and reaches around 12 million square kms (the warming of this pool also leads to rise in sea levels to the extent that Western Equatorial Pacific surface is 30 cm higher than Eastern Equatorial Pacific surface) in a time span of 2–7 years. As the warm humid air rises from this pool, precipitation occurs in the region that extends up to Indonesia (Walker Circulation). During the movement of warm surface water towards west of Pacific Ocean, cool water is drawn to the surface in the East Pacific, a process called 'upwelling' that lifts nutrients from the dark ocean beds to the surface. These nutrients are consumed by planktonic algae providing one of the richest sources of proteins to Peruvian fishery.

However, as the surface temperature of this pool of warm surface water exceeds 26.5 °C, cyclone formation takes place sapping the trade wind energy and ultimately dismantling *warm surface water pool*. The net result is reversal of warm water towards East Pacific causing El Nino conditions. During this period, the trade winds settle down in the central and western Pacific, allowing warm water to accumulate in the surface, causing the nutrients produced by the upwelling of cold water to significantly come down, leading to the killing of plankton and other aquatic life such as fish and the starvation of many seabirds.

This change in wind direction and speed makes the atmosphere act as a huge blanket trying to obstruct the rotation of the Earth. Research conducted by American Geophysical Union and NASA estimates that the Earth's rotation slows down by an average of 300–400 μs as a result of El Nino making days longer. Some other phenomenon occurs concurrently with El Nino including an increase in seismic activity along the East Pacific Rise (a submarine mountainous range) and a change in sea level across the Pacific basin indicating a shift in the regional gravity field and their link to planetary waves. El Nino is also preceded by abnormal temperature rise in the ocean waters surrounding the Indonesian Archipelago. A wide variety of disasters have been blamed on the El Nino effect including a famine in Indonesia in 1983, bush fires in Australia arising from droughts, rainstorms in California and the destruction of anchovy fishery off the coast of Peru. During 1982/83, it is said to have led to the death of some 2000 people worldwide and caused losses amounting to approximately 12 billion dollars. In 1997–98, many nations were afflicted with violent swings in weather pattern. Australia was severely affected by drought, Chinese coastal provinces were affected by deadly storms, large-scale fires devastated Indonesian forests caused primarily to due severe dry conditions in last 50 years, tornadoes tore apart parts of USA, chronically drought-affected Ethiopia was flooded with water, and large-scale forests fires destroyed rain forests of South America.

La Nina (pronounced as Lah Nee-Nyah), on the other hand, is the female sibling of El Nino and is distinguished from El Nino by abnormally cold ocean temperatures in the Equatorial Pacific region in contrast to that of El Nino when the ocean temperatures in the same region are unusually warm. As a result, the impacts of La Nina on the global climate and ocean temperature tend to be opposite those of El Nino. In the USA, winter temperatures are warmer than normal in the south-east and cooler than normal in the north-west during a La Nina year. Snow and rain are experienced on the west coast, and unusually cold weather is observed in Alaska.

It has also been seen that there are higher than normal number of hurricanes in the Atlantic during La Nina year.

References

Altomonte, Sergio. 2008. *Climate change and architecture: Mitigation and adaptation strategies for a sustainable development.*

Barkov, N.I., E.S. Korotkevich, F.G. Gordienko, and V.M. Kotlyakov. 1977. The isotope analysis of ice cores from Vostok station (Antarctica), to the depth of 950 m. *IAHS-AISH Pub* 118: 382–397.

Donald, R.A.P.P. 2014. *Assessing climate change: Temperatures, solar radiation and heat balance.* Berlin: Springer.

Fortenbaugh, William, Pamela Huby, Robert Sharples, and Dimitri Gutas (eds.) 2016. *Theophrastus of eresus. Sources for his life, writings, thought and influence (2 Vols)*, vol. 54. Netherlands: Brill.

Glick, Daniel. 2004. The big thaw. *National Geographic* 18.

Herman, J.R., G.J. Labow, W. Qin, L. Huang, M.T. DeLand, J. Mao, S. A. Lloyd, and D. Larko. 2012. A net decrease in the earth's cloud plus aerosol reflectivity during the past 33 years (1979–2011). In *AGU fall meeting abstracts*, vol 1, 3.

IPCC. 2007. IPCC fourth assessment report: Climate change 2007 synthesis report. Valencia.

Jansen, E., J. Overpeck, K.R. Briffa, J.-C. Duplessy, F. Joos, V. Masson-Delmotte, D. Olago, B. Otto-Bliesner, W.R. Peltier, S. Rahmstorf, R. Ramesh, D. Raynaud, D. Rind, O. Solomina, R. Villalba and D. Zhang. 2007. Palaeoclimate. In *Climate change 2007: The physical science basis. Contribution of Working Group I to the fourth assessment report of the intergovernmental panel on climate change*, ed. Solomon, S., D. Qin, M. Manning, Z. Chen, M. Marquis, K.B. Averyt, M. Tignor and H.L. Miller. Cambridge University Press, Cambridge, United Kingdom and New York, NY, USA.

Kunzig, Robert, and I. Block. 2011. World without ice. *National Geographic* 220: 90–109.

Le Treut, Hervé, Richard Somerville, Ulrich Cubasch, Yihui Ding, Cecilie Mauritzen, Abdalah Mokssit, Thomas Peterson, and Michael Prather. 2007. Historical overview of climate change.

Loutre, M.F. 2003. Ice ages (Milankovitch theory). *Encyclopedia of Atmospheric Sciences* 995–1003.

Maharjan, Keshav Lall, and Niraj Prakash Joshi. 2013. *Climate change, agriculture and rural livelihoods in developing countries.* Berlin: Springer.

McElwain, Jennifer C., Peter J. Wagner, and Stephen P. Hesselbo. 2009. Fossil plant relative abundances indicate sudden loss of Late Triassic biodiversity in East Greenland. *Science* 324 (5934): 1554–1556.

Met Office. 2015. The Great Smog of 1952. http://www.metoffice.gov.uk/learning/learn-about-the-weather/weather-phenomena/case-studies/great-smog.

National Research Council. 2011. *Understanding earth's deep past: Lessons for our climate future.* National academies Press.

Schaub, Georg, and Thomas Turek. 2011. *Energy flows, material cycles and global development.* Berlin: Springer.

Steinthorsdottir M, and Vajda V. Early Jurassic (late Pliensbachian). 2015. CO_2 concentrations based on stomatal analysis of fossil conifer leaves from eastern Australia. Gondwana Research. 27(3):932–939.

Wang, Lawrence K., and Chih Ted Yang (eds.). 2014. *Modern water resources engineering.* New York: Humana Press.

Chapter 3
Understanding the Warming Process

> *The more we convert carbon into carbon dioxide, the more we deplete oxygen in the atmosphere.*

Abstract Greenhouse gases form a thick blanket around the Earth to regulate the loss of energy into space, thereby keeping the Earth warm enough for the survival of living beings. However, during the past three centuries the uncontrolled and unregulated emission of carbon dioxide, methane and nitrous oxide, ozone and other long-lived industrial gases such as CFCs, HFCs and PFCs have changed the global energy balance. This has resulted in warming of the Earth, and the current estimates are that a temperature increase of 4 °C may lead to disastrous and irreversible changes. Carbon dioxide and nitrogen oxides are the largest source of GHGs that are being released due to deforestation, land diversion, burning of coal and petroleum, soil erosion and water pollution. As a consequence, large quantities of carbon stored in living and dead vegetation, soil organic matter, as dissolved organic and inorganic carbon in oceans, have been liberated into the atmosphere as carbon dioxide or methane with serious impact on global climate. Increase in nitrogen in the atmosphere due to extensive use of fertilizers and combustion processes, loss of blue-green algae and nitrogen-fixing bacteria has further aggravated the warming process.

Keywords Greenhouse gas · Biological pump · Biological calcification
Biological accumulation ratio · Carbon · Nitrogen · Methane
Dissolved inorganic carbon (DIC)

3.1 The Greenhouse Gases and Their Effects

Of the 3.9×10^{26} J of energy radiated by the Sun each second, what is intercepted by Earth is only 1.75×10^{17} J or 1.52×10^{18} kilowatt per year in the form of electromagnetic radiation (Abdel-Motalib & Allam 2009). A substantial portion of this energy is reflected back by the Earth's atmosphere into the space, and the balance is retained in the atmosphere, clouds, land surface and ocean as heat energy (infrared). The clouds, aerosols and atmosphere reradiate some energy back to Earth, a phenomenon called **natural greenhouse effect**. This effect is responsible for keeping the Earth warm by nearly 30 °C for the survival and growth of living beings. There are six main gases (Table 3.1) responsible for greenhouse effect, and these are:

- Carbon dioxide
- Methane
- Nitrous oxide
- Ozone
- Water vapour
- Halocarbons

The spectacular phenomenon of greenhouse effect was first noticed in 1681 by Edme Mariotte (Prichard 2010) who found that while the Sun's light and heat could pass through the glass screen, heat from other sources could not. Later on, Joseph Fourier claimed in 1824 that the temperature of the Earth can be increased by the interposition of the atmosphere because heat in a state of light finds less resistance in penetrating the air than in repassing into the air in the form of non-luminous heat. In 1859, John Tyndall suggested that changes in the amount of any of the radiatively active constituents of the atmosphere such as water and CO_2 could have produced all the mutations of climate (Hulme 2009). Callendar (1938) found that a 100% increase in the atmospheric CO_2 concentration resulted in 20 °C increase in global mean temperature (Hawkins and Jones 2013). Initial understanding of the greenhouse effect was restricted to only water and carbon dioxide which were considered as GHG responsible for preventing infrared wave from escaping Earth's atmosphere and redirecting them to Earth. It was only after 1970 that the natural processes of interactions between atmospheric gases, Earth and ocean were better understood and the contribution of other gases in greenhouse effect was recognized.

During the past three centuries, the emission of carbon dioxide (mainly due to burning of coal, oil, natural gas and deforestation), methane and nitrous oxide (primarily due to changes in land use and agriculture), ozone (due to automobile exhaust and other sources) and other long-lived industrial gases such as CFCs, HFCs and PFCs has changed the global energy balance (Fig. 3.1). Thick blanket of greenhouse gases restricts the loss of energy to space, thereby keeping the Earth warmer than before. The composition of the atmospheric constituents is a dynamic phenomenon determined by natural and anthropogenic emission of gases and aerosols, their transport at a variety of scales, chemical transformation and exchange by the terrestrial and aquatic ecosystems (Figs. 3.2 and 3.3).

3.1 The Greenhouse Gases and Their Effects

Table 3.1 Main greenhouse gases

Source	Gases		
Fossil fuels, forest clearance and forest burning	Carbon dioxide	1. Lasts up to 100 yrs in the atmosphere; 2. Most abundant heat absorber; 3. Accounts for 50% of man-made share of global warming problems; 4. Concentration rising with burning of fossil fuel and forests	Annual emissions have grown from 21 to 38 gigatons between 1970 and 2004; Global atmospheric concentration increased from a pre-industrial value of approximately 280–379 ppm in 2005; Analysis of ice core shows that CO_2 concentration fluctuated between 180 and 300 ppm over the glacial–interglacial cycles of the last 650,000 years
Agricultural activities, fossil fuel, bovines, change in land use and other activities	Methane	1. Lasts up to 10 years in atmosphere; 2. Absorbs 20–30 times more heat than CO_2; 3. Generated by bacteria which break down organic matter under anaerobic conditions	Its global atmospheric concentration increased from a pre-industrial value of approximately 715 parts per billion to 1774 parts per billion in 2005
Agricultural activities, fossil fuel, change in land use and other activities	Nitrous oxides	1. Lasts up to 180 years; 2. Produced by microbes in soil; 3. Absorbs 200 times more heat than CO_2 4. Increase due to chemical fertilizer, slash and burn and fossil fuel emission	The global atmospheric concentration increased from a pre-industrial value of about 270 parts per billion to 319 parts per billion in 2005
Halocarbons	CFC, HFC, PFC	1. Lasts up to 400 years; 2. 16,000 times more effective than CO_2 in absorbing heat	Pre-industrial concentration was zero

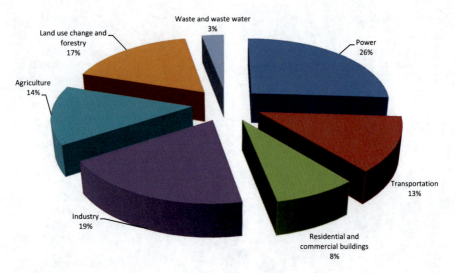

Fig. 3.1 Sources of greenhouse gas emission (*Data Source* IPCC 2007)

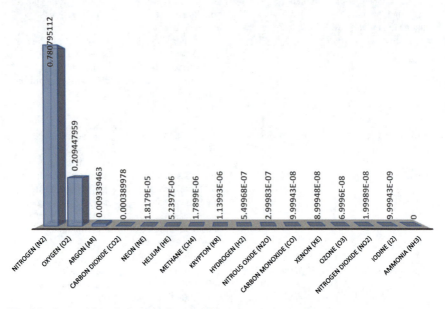

Fig. 3.2 Composition of atmospheric gases (%)

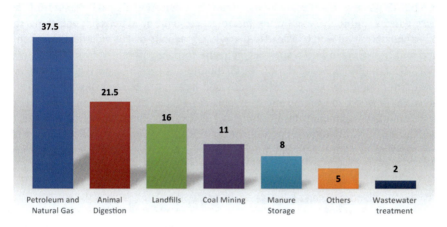

Fig. 3.3 Per cent methane emission by source in the USA (*Data Source* Stone 2014)

3.2 The Carbon Factor

More Carbon—More Prosperity

More Prosperity—More Consumption

More Consumption—More Carbon

The word 'carbon' has been derived from the Latin word 'carbo' which means coal. Carbon is the fourth most abundant element on this planet after hydrogen, helium and oxygen. Most of the carbon is stored (Fig. 3.4) in various forms in the Earth as coal and hydrocarbon and in the atmosphere as CO_2 and black carbon.

How coal was formed is still an enigma for the scientific world. There are two important components of a tree, cellulose which is an ancient material composed of

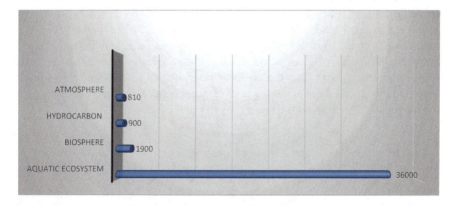

Fig. 3.4 Carbon (in gigaton) in various systems of the Earth

long chains of sugar molecules and forms the walls of plant cells and lignin, which is made of phenols, a molecule hard to digest by micro-organisms. It is believed that during the 60 million years of Carboniferous period the dead trees fell one above the other and the buried wood, instead of decaying, gradually lost its volatile components and was transformed into a substance closure to pure carbon (Nelsen et al. 2016). Why most of this happened during the Carboniferous period? One hypothesis is that the Earth during the period was swampy and lacked oxygen that slowed down the activities of wood-consuming micro-organisms. Another hypothesis suggests that the coal formation was mainly due to geological factors rather than biological. It suggests that during the Carboniferous period, the continents were moving and colliding with one another forming basins and mountains which buried sediments laden with dead and dying trees for millions of years thereafter. Whatever is the reason, the fact remains that the accidental discovery and use of coal during Industrial Revolution changed the landscape and atmosphere of the world.

The earliest reference to the use of coal as fuel by humans is from the geological treatise '*On Stones*' by Theophrastus where he mentions two types of coal, one which retains its heat and can be rekindled by fanning and the other that was ignited by the heat of the Sun, especially when sprinkled with water. Englishmen started using coal for fuel as early as ninth century which was subsequently banned during the reign of King Edward I (1272–1307). Richard II (1377–1399) revoked the ban and introduced taxation. This was succeeded by strict regulatory measures for use of coal by Henry V (1413–1422). Unfortunately, by sixteenth century, much of the natural forests in England were cleared to meet the growing requirement of fuel, timber, shipbuilding and farmland. Englishmen were left with no option but to use coal as a substitute for wood energy.

With the onset of Industrial Revolution in eighteenth century, the consumption of coal shot up rapidly and it replaced wood as a major source of fuel for transportation, industries, household energy and electricity generation. Currently, coal is responsible for 40% of the global energy needs, is the second best source of primary energy after oil and tops the list of sources of electricity generation. Black carbon is a component of soot that remains in the atmosphere only for a few days or weeks. It is a product of incomplete combustion of fossil fuels, biofuels and biomass originating from burning of biomass, diesel vehicles, residential buildings, etc. (Fig. 3.5), using solid fuels such as coal, wood, dung and agriculture residues. Black carbon (Fig. 3.6) contributes to global warming by directly absorbing solar radiations and indirectly by darkening the surface of ice and snow. Hydrocarbons, whether oil or gas is generated in the sedimentary rocks from the organic matter of dead animals and plants that is deposited along with the sediments and the physiochemical changes of thermogenic and biogenic origin, convert the organic matter into oil/gas.

> **Box 3.1—Environmental Effects of Burning Coal**
> A 500 MW thermal plant burns 1,430,000 tons of coal and uses 2.2 billion gallons of water and 146,000 tons of limestone to produce 3.5 billion kilowatt-

3.2 The Carbon Factor

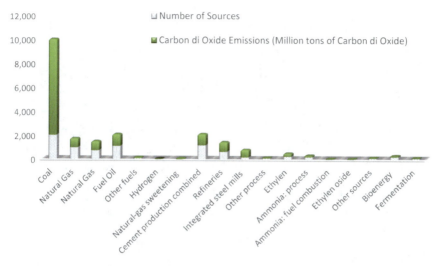

Fig. 3.5 Profile of global carbon dioxide stationary sources emitting more than 0.1 million tons of carbon dioxide per year (*Source* IEA GHG 2002)

---

hours per year (Ma 2011). The electricity so produced is sufficient for a city of about 140,000 people.

In addition to generating electricity, the thermal power plant also releases:
- *10,000 tons of sulphur dioxide*—which is the main cause of acid rain and damages forests, lakes and buildings;
- *10,200 tons of nitrogen oxide*—which is the major cause of smog and also a cause of acid rain;
- *3.7 million tons of carbon dioxide*—the main greenhouse gas, and is the leading cause of global warming;
- *500 tons of small particles*—small particulates are a health hazard, causing lung damage;
- *220 tons of hydrocarbons*—fossil fuels are made of hydrocarbons, and all partially burnt or incompletely burnt hydrocarbons are released into the air causing smog;
- *720 tons of carbon monoxide*—a highly poisonous gas for humans;
- *125,000 tons of ash and 193,000 tons of sludge from the smokestack scrubber*—the ash and sludge consist of coal ash, limestone and many pollutants, such as toxic metals like lead and mercury;
- *225 lb of arsenic, 114 lb of lead, 4 lb of cadmium and many other toxic heavy metals; and*
- *Trace elements of uranium*—all but 16 of the 92 naturally occurring elements have been detected in coal, mostly as trace elements below 0.1% (1000 parts per million, or ppm).

(References: Koroneos and Papadopoulos 2010 and Reddy 2013)

The inefficient burning of wood based fuel, forest fires, burning of agriculture waste, coal and diesel produce large quantity of soot[1] which is a powerful absorbent of solar radiation thereby heating atmospheric temperature. Scientific studies have indicated that high levels of soot in the atmosphere are responsible for glacial melting in the Himalayas and changes in monsoon circulation.

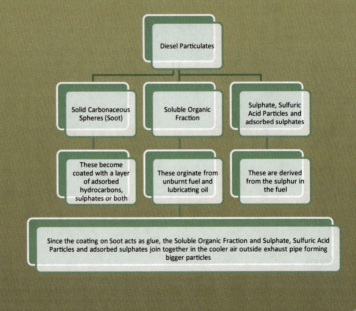

Fig. 3.6 Black carbon—the by-product of inefficient burning (*Source of Data* Srivastav 2016)

3.2 The Carbon Factor   47

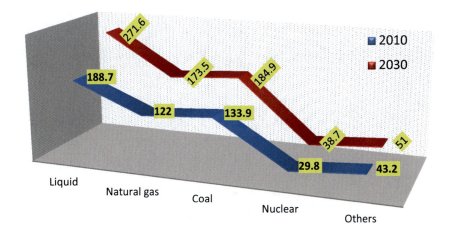

Fig. 3.7 World energy consumption (quadrillion BTU) (*Source* EIA 2007, Appendix E)

If the current trend in population growth and slow economic development continues, chances are that the reliance of many developing nations in Asia and Africa on non-renewable energy (Fig. 3.7) such as primarily coal and gas will continue irrespective of the fact whether new and efficient alternatives are available or not. For example, China will increase its consumption of coal from 55.6 quadrillion Btu in 2010 to 95.7 quadrillion Btu in 2030 (Sieminski 2014). Similarly, India's consumption of coal will increase substantially from 9.1 quadrillion Btu in 2010 to 14.3 quadrillion Btu in 2030 (Sieminski 2014). There will, definitely, be consequences, and the extent of damage will be felt by the future generations.

3.2.1 Data set on Carbon and Carbon Dioxide

Data provided in Figs. 3.8, 3.9, 3.10, 3.11, 3.12, 3.13, 3.14, 3.15 and 3.16 is intended to provide an understanding of the changes in carbon and carbon dioxide content in the atmosphere since the onset of Industrial Revolution . The molecular weight of carbon is 12 and that of carbon dioxide is 44. Therefore, each ton of carbon releases 3.67 tons of carbon dioxide in the atmosphere.

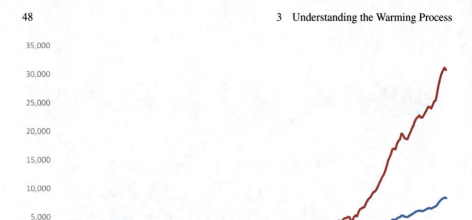

Fig. 3.8 Carbon and carbon dioxide added to atmosphere between 1751 and 2007 (*Source* EIA 2007)

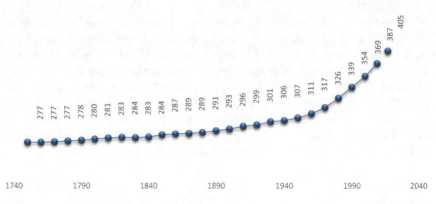

Fig. 3.9 Decadal changes in carbon dioxide concentration (parts per million by volume) between 1750 and 2017 (*Source* EIA 2007 and Blunden et al. 2018)

3.2.2 Properties of Carbon Dioxide

In nature, carbon dioxide[1] exists as a weak acidic gas which reacts easily with the alkaline sea water to form carbonic acid. Carbonic acid in turn dissociates into bicarbonate ion (HCO_3^-), carbonate ion (CO_3^{2-}) and hydronium ion (H^+). The reaction between atmospheric CO_2 and surface ocean water is determined by the chemical equilibrium between CO_2 and carbonic acid in sea water, the partial pressure of CO_2 in the atmosphere and the rate of air/sea exchange. The sum of carbon

---

[1] Carbon dioxide has a special affinity to absorb infrared radiations.

3.2 The Carbon Factor

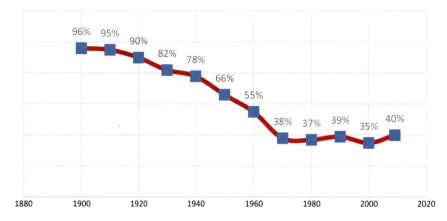

Fig. 3.10 Temporal change (%) in release of carbon dioxide—by burning of coal versus total carbon dioxide released by burning coal, oil and natural gas (*Source of Data* EIA 2007)

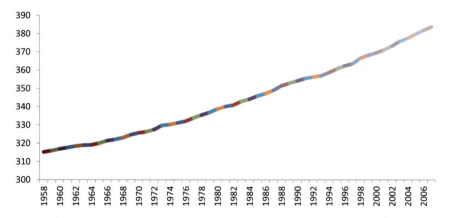

Fig. 3.11 Annual mean carbon dioxide concentration (ppmv) measured at Mauna Loa (*Source of Data* ESRL 2016 ftp://aftp.cmdl.noaa.gov/ccg/co2/trends/co2_mm_mlo.txt)

contained in carbonic acid, carbonate ion and bicarbonate ion is known as total dissolved inorganic carbon (DIC) (Rubin and Coninck 2004).

Sir Charles David Keeling was a scientist from San Diego who developed an accurate technique for measuring carbon dioxide levels in the atmosphere (Weart 2008). His initial observations in 1957 revealed 310 ppm (parts per million by volume) of carbon dioxide. Keeling's subsequent measurements also showed a rapid increase in carbon dioxide level and an enormous disruption in the global carbon cycle. At the time of his death in 2005, the atmospheric carbon dioxide level was 380 ppm which went up to 391 ppm in 2011 (Weart 2008). Subsequent researches have established that not only carbon dioxide but the levels of other greenhouse gases have also risen dramatically since the last century. Carbon dioxide levels have increased by about

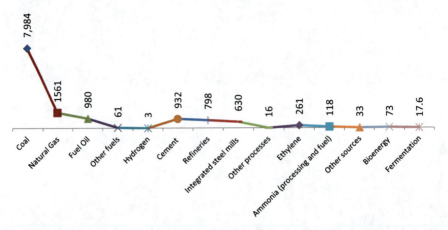

Fig. 3.12 Profile of CO_2 from stationary sources emitting more than 0.1 Mt CO_2 per year (*Source of Data* IEA GHG 2002)

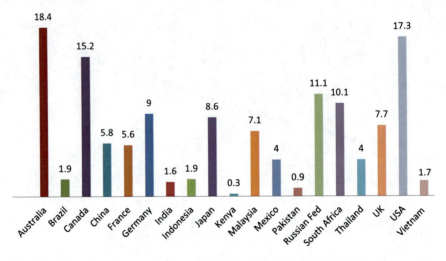

Fig. 3.13 CO_2 emission per capita in metric ton (*Data Source* World Bank 2006)

25% since 1850, methane levels by 100% and nitrous oxide levels by 15% (Schneider 1989).

Both CO_2 (carbon dioxide) and CH_4 (methane) play a vital role in the natural cycle of carbon. Terrestrial plants capture carbon dioxide from the atmosphere, convert it into carbohydrate and store it in plant body and in soil. Organic carbon compound stored in the plants and soil is consumed at different rates depending on the rate of plant respiration and soil microbial activity which in turn depends on soil temperature and moisture. Respiration by plants and their decomposition returns carbon to the atmosphere either as carbon dioxide (under aerobic conditions) or as CH_4 (under anaerobic conditions). In addition to respiration by plants and their decomposition,

3.2 The Carbon Factor

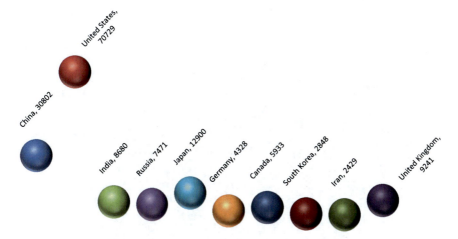

Fig. 3.14 Total carbon dioxide released by top ten emitters (million tons of carbon) (*Source of Data* EIA 2007)

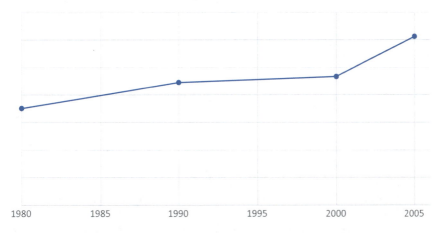

Fig. 3.15 World coal consumption (Quadrillion BTU) (*Data Source* EIA 2010)

biomass burning also adds a significant quantity of carbon dioxide and CH_4 in the atmosphere. Changes in rate of photosynthesis in terrestrial ecosystems in response to changes in carbon dioxide concentration, temperature variation, precipitation and nutrients availability also affect carbon dioxide release.

Prior to the onset of Industrial Revolution (in 1750), the atmospheric concentration of carbon dioxide was relatively stable and varied between 260 and 280 ppm for nearly ten thousand years (Watson et al. 1990). After 1750, the carbon dioxide concentration has risen, at an unprecedented rate touching 380 ppm in 2005 and 391 ppm in 2011 (Solomon et al. 2007). The average annual increment in the atmospheric carbon dioxide has been 3.2 ± 0.1 GtC per year in the 1990s to 4.1 ± 0.1 GtC per year in the

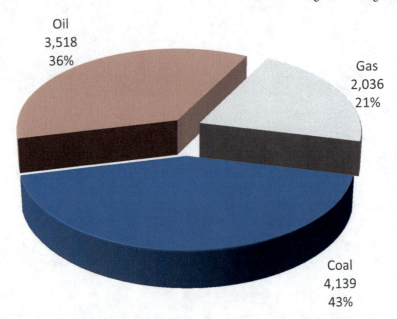

Fig. 3.16 Global projection of carbon dioxide emissions from fossil fuel (million tons of carbon) combustion in 2020 (*Source* EIA 2007)

period 2000–2005 (Solomon et al. 2007). Most of the additional (post-1750) carbon dioxide termed as 'anthropogenic carbon dioxide' has been contributed (Fig. 3.17) by the following sources:

1. **Carbon dioxide from fossil fuel burning and cement production**[2]—Carbon dioxide release from fossil fuel burning and cement manufacturing has continuously risen from 5.4 ± 0.3 GtC/yr in the 1980s to 6.4 ± 0.4 GtC/yr in the 1990s and further to 7.2 ± 0.3 GtC/yr during the period 2000 to 2005 (Raupach et al. 2007).
2. **Carbon dioxide from deforestation and agricultural development**—Carbon dioxide stored for decades to centuries and more in the living biomass and soil has been released in large quantities in recent years. Most of the deforestation and land conversion during the twentieth century have occurred in the tropical regions as countries in mid-latitude had already eliminated their old growth forest in the nineteenth and early twentieth century. Deforestation, degradation and land use change have injected carbon @ 2 ± 0.8 GtC/yr in the 1980s to 2.2 ± 0.8 GtC per year in the 1990s (Watson et al. 1992).

More than 75% increase in carbon dioxide concentration during post-industrial era is due to fossil fuel combustion and cement manufacturing. Balance 25% comes

---

[2]Carbon dioxide in the Earth's atmosphere absorbs and emits infrared radiation at wavelengths of 4.26 μm (asymmetric stretching vibrational mode) and 14.99 μm (bending vibrational mode), thereby playing a crucial role in the greenhouse effect.

3.2 The Carbon Factor

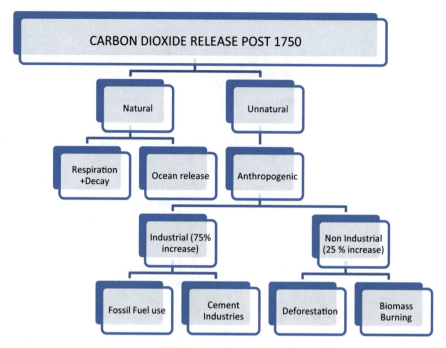

Fig. 3.17 Carbon dioxide release post-1750

from deforestation, forest fire, biomass burning and land use changes. Excess carbon produced due to human-induced factors is partially offset by processes such as photosynthesis (land-based sink) and ocean uptake (ocean-based sink) that act as natural sinks of carbon and absorb nearly 3.3 gigaton of carbon per year (Pachauri and Reisinger 2007). Of the total quantity emitted, oceans are estimated to have taken 30% (118 ± 19 GtC), terrestrial ecosystems have taken up around 25% through reforestation and revegetation and 45% remains in the atmosphere (Pachauri and Reisinger 2007).

As a consequence of increase in atmospheric carbon dioxide, the ratio of oxygen to nitrogen has declined for the reason that oxygen is used when the fossil fuels are burned. Moreover, it has also been observed that the relative amount of Carbon-13 isotope[3] in the atmosphere, which is more abundant in the oceans, and geothermal or volcanic eruptions, has been declining, an indication that most of the additional carbon comes from fossil fuel and biomass burning.

---

[3] Carbon-13 isotope is a heavy form of carbon that has been found to be less abundant in vegetation and fossil fuels but is more abundant in carbon found in volcanic and geothermal emissions. There is another rare radioactive isotope of carbon called Carbon-14 which is present in atmospheric carbon dioxide but absent in fossil fuels.

3.2.3 Carbon in the Terrestrial System

A dragon of fire, running wild through the forest, Setting every tree ablaze and touching off bright torches.
Like a monster starved for a century, Destroying everything;
Like a rocket, the fire raced everywhere, Giving vent to rage—rage pent up for thousands of years

The terrestrial system of the Earth has been divided into 14 major biomes that cover nearly 149 million square km. Forests, the major carbon sink, currently occupy around 40 million sq km (Table 3.2) and are subdivided into primary (36%), modified primary (53%), semi-natural (7%) and plantation forest (4%). This figure of 40 million square km may not be accurate and reliable as global estimates of forest cover and forest cover change are difficult because of conflicting definitions of forest, lack of satellite and radar data, authenticity of country past and present records and unmonitored land use change.

These forest lands are unevenly distributed with five (Russian Federation, Brazil, Canada, USA and China) out of 229 countries accounting for more than half of total forest area (2097 million hectare). The remaining 47% is spread among 217 countries of which 64 countries have less than 10% of their land covered under forest (Srivastav and Srivastav 2015). The skewed distribution of forests has been largely responsible for the destruction of pristine or primary forests.

Over exploitation of natural forests for direct human consumption (commercial[4] and non-commercial) is just one of the many factors that have contributed to the loss and degradation of forests and release of a large quantity of carbon dioxide in a short span. The expansion of agriculture, human settlements and infrastructure development are other major contributors to global climate change. It is worth noticing that the area under agriculture and other crops expanded from 426 million hectare in 1980 to 453 million hectares in 1995, largely at the expense of forest cover which decreased by 42.6 million hectare over the same period (UNEP 2013). During the

Table 3.2 Forest area by category

Type	Area (million sq km)
Primary	14.56
Modified natural/primary	21.08
Semi-natural	2.84
Forest plantation productive	1.2
Forest plantation protective	0.32
Total	40

Data Source FAO (2006)

---

[4]High global demand for timber by industrialized countries during the second half of twentieth century leads to heavy loss of forests in South America, Africa and Asia. Africa lost more than 52 million hectares in the 1990s alone.

1990s, the total loss of natural forests was 16.1 million hectares each year mainly due to commercial logging, conversion to permanent agriculture and urbanization. This conversion of forests to agricultural land and other activities still continues at an alarmingly high rate leaving most of the biomes fragmented (UNEP 2013).

Let us understand how carbon and oxygen add value to the soil and vegetation. All green plants containing chlorophyll whether in the forests or outside in the form of agriculture crop or otherwise absorb solar energy and carbon dioxide to produce biomass for growth and reproduction in the following manner:

1. *Solar energy ± plant = Gross Primary Productivity (carbon dioxide is accumulated and stored);*
2. *Gross Primary Productivity*[5]—*Respiration* **(carbon dioxide is released by plants)** *= Net Primary Productivity; and*
3. Gross Primary Productivity of a unit area—Respiration by plants, animals and saprophytes = **Standing crop or accumulated biomass.**

It is interesting to note that only 2% of solar energy is absorbed by plants and this energy is sufficient to feed billions of humans and trillions (rather zillions) of animals on Earth.

Since respiration is a continuous process, Gross Primary Productivity (of a plant or all plants in a given area) is a dynamic process whereby plants accumulate biomass and use part of biomass for respiration. As a general principle, most of the energy built up during Gross Primary Productivity is utilized for metabolic activities and only 20% of the GPP energy goes in the standing crop of trees, plants and animals.

It is also interesting to understand that during the productivity phase of the plants, the **biomass accumulation ratio** (Fig. 3.18) in young fast-growing plants is low as compared to old growth forests because most of the energy in young plants is used for growth.

$$\text{Biomass Accumulation Ratio} = \frac{\text{Standing Crop}}{\text{Net Primary Productivity}}$$

In the year 2005, the total growing stock of the global forest was estimated at 434 billion cubic metre and it has been estimated that this growing stock of world's forests stored 283 gigaton of carbon (Fig. 3.19) in their biomass and 638 gigaton of carbon in the terrestrial ecosystem as a whole (up to a soil depth of 30 cm) (Schneider 2004).

IPCC estimated that the average carbon stock (Fig. 3.20) of world's forests during 1990, the mid-1990s and 2005 was 82 tons per hectare, 86 tons per hectare and 81 tons per hectare, respectively (Global forest resources assessment 2005).

One cubic metre of growing stock in a forest area is equal to 1.3 ton of total biomass.[6] This 1.3 ton of biomass contains 0.7 ton of carbon and 0.6 ton of other elements. In other words, one cubic metre of growing stock in a forest sequesters

---

[5]GPP can be calculated for a plant or all plants in a given area.

[6]One cubic metre of growing stock = 1.3 ton of total biomass = 1 ton of above ground biomass + 0.3 ton of below ground biomass = 0.7 ton of carbon + 0.6 ton of other elements.

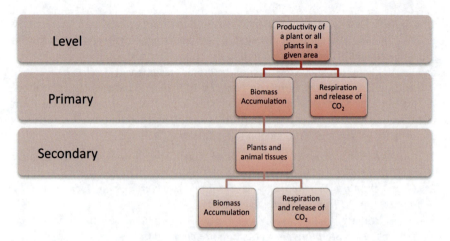

Fig. 3.18 Biomass accumulation in plants

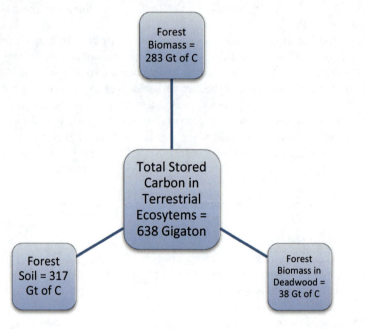

Fig. 3.19 Illustration of stored carbon in terrestrial ecosystem (*Data Source* Schneider 2004)

0.7 ton of carbon. Until we reverse the current trend of deforestation and land use change, the carbon stored in these forests will be released into atmosphere further exacerbating the global warming.

In addition to the commercial logging and conversions, the forests have been significantly damaged by extreme weather events such as fire, droughts and cyclones.

3.2 The Carbon Factor

Fig. 3.20 Average carbon stock per hectare (tons) (*Source Data* Global forest resources assessment 2005)

Forest fires are a serious problem in the southern countries along the rim of the Mediterranean (Croatia, France, Italy, Greece, Slovenia and Spain). Fire-prone area in North American forests has grown since the 1970s mainly due to fuel build-up from past fire protection programmes (Schoennagel et al. 2017). Russian Federation lost large areas of forests to insect attack (46%), fire (33%) and extreme weather events (16%) in the mid-1990s (van Lierop et al. 2015). Deforestation is also linked to severe flood events and landslides. At the French–German border, the floodwaters of Rhine rose more than 7 m above flood level once every twenty years between 1900 and 1977 (Abramovitz 2001). Scores of countries continue to suffer from disastrous flooding as a result of deforestation and soil loss. For instance, the Yangtze River basin experienced one of the worst flooding in 1998 (Brown 2013) after losing 85% of its original tree cover. Similarly, Mozambique was partially inundated as the Limpopo River over flooded its banks in 2000 killing thousands of people and destroying homes and crops at an unprecedented scale. The main cause of this disaster was a loss of 99% natural forest cover in the Limpopo River basin (Brown 2013).

A 'business-as-usual' scenario suggests that continued rapid economic growth and industrialization may result in further environmental damage and that most of the developing and underdeveloped regions may become more degraded, less forested, more polluted and less ecologically diverse in future. Continued logging in Brazilian Amazon, the Congo basin and Borneo will further accelerate the frequency of natural hazards making humanity more vulnerable to natural events. The 190 million hectare Africa's Congo basin, the world's second largest rainforest, spanning ten countries and protecting 400 species of mammals including world's largest population of Gorillas, bonobos, chimpanzees and elephants, is shrinking by 1.6 million hectare a year (Brown 2008). China's logging spree in Indonesia, Myanmar, Papua New Guinea and Siberia to meet the explosive global demand of wood products will soon decimate a substantial chunk of primary forests. Similarly, in Brazil some 58 million hectares of land are affected by desertification (Brown 2008).

In addition to deforestation and fire, use of forests for obtaining fuel-wood and charcoal, the oldest form of energy source, continues to persist is almost all countries

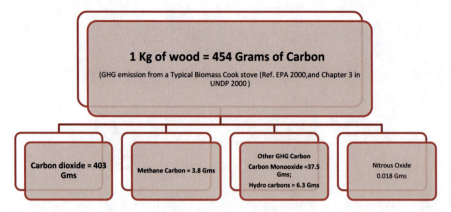

Fig. 3.21 GHG emission from biomass-based cook stove

and is still the predominant form of wood energy in rural areas of most developing countries. Wood for fuel primarily comes from the following sources:

i. As by-product of logging operations;
ii. By felling trees with minimal timber value;
iii. By lopping and pollarding standing trees;
iv. By collecting dead, drifted and fallen wood; and
v. As residue from sawmills and other processing industries.

Currently, the wood energy accounts for nearly 5% of world's total energy supply which ranges from 1% in Europe to 75% in Ethiopia and Congo. Seventy-two percentage of the total rural households and 33% of the total urban households in India use wood as fuel (Agarwal et al. 1999). Over half of wood fuel people live in China, India and Indonesia and wood fuel, dung and crop residue will continue to remain in great demand for more than 2.5 billion people in developing countries especially those living in and around the forests and protected areas (IEA 2010) (Fig. 3.21).

The reasons for this heavy reliance are not difficult to understand. Firstly, supply of wood is assured vis-a-vis energy from thermal, nuclear, oil, hydro and other sources. Secondly, the cost, if any, is within the paying capacity of local population, and lastly, the sociocultural factors favour use of wood fuel for cooking food rather than using coal or LPG. To a large part of the developing world where commercial sources of energy are not a viable alternative, wood fuel is perceived as a blessing in disguise that keeps the energy issue away from crisis situation. It is for this reason, probably, that many nations intentionally retain wood fuel in non-commercial sector for fear of severe backlash by local communities. In a wood fuel study in a Ghana village, it was found that with increase in price of wood fuel, the household budget for wood fuel purchase rose from 1 to 16.3% (Mohammed and Osei-Fosu 2016). However, this rise impacted the budget for health and education as money was diverted for wood fuel and people resorted to cooking fewer meals or consumed uncooked food (Table 3.3).

3.2 The Carbon Factor

Table 3.3 People relying on biomass fuels for their energy needs

	Total population (million)	Rural (million)	Urban (million)
Sub-Saharan Africa	575	413	162
North Africa	4.2	4	0.2
India	740	663	77
China	480	428	52
Indonesia	156	110	46
Rest of Asia	547	455	92
Brazil	24	16	8

Source IEA (2006)

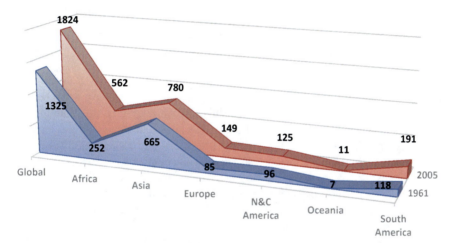

Fig. 3.22 Increase in wood fuel production (million cubic metres) (*Data* FAO 2005)

Analysis based on the available data indicates that the average per capita consumption of wood fuel is around 0.7 kg/day. If urban population is excluded, the per capita share goes up by more than three times (2.5 kg/day). Globally, there has been an increase of about 500 million cubic metres in the last 50 years and the same is expected to grow further since it is directly linked to both population growth and income level, with low incomes generally indicating higher consumption. Wood fuel burning and its impact on carbon dioxide release will pose a gigantic problem in the coming future, the result of which cannot be predicted with certainty (Figs. 3.22 and 3.23).

The general presumption that conservation is an impediment for economic growth has, ironically, changed the original condition of most natural ecosystems in composition, extent and quality. Areas devoid of original forest ecosystems have lost the ability to sequester carbon, stabilize soil, recycle nutrients and recharge water tables. Natural regeneration, forest planting through single or multiple species and

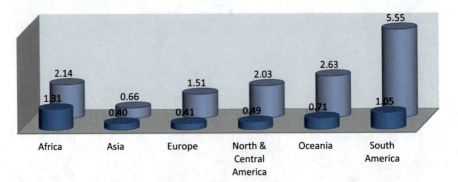

Fig. 3.23 Global consumption pattern of wood fuel (Data FAO 2005)

landscape restoration may provide environmental relief, but they cannot substitute the ecological and protective value of pristine forests. As tree cutting, hunting and diversion in many parts of the world accelerate, the naturalness and the integrity of nearly half of these ecosystems are at risk. The World Resources Institute has estimated that almost 40% of the world's natural and pristine forests will be compromised in next decade or two if the current deforestation rates continue. Since 1960, global industrial wood production has risen by 50%, to 1.5 billion cubic metres, four-fifths of which comes from primary and secondary growth forests (Brown et al. 2013). About the same quantity, 1.8 billion cubic metres, is burned directly as wood fuel each year. Each kilo of wood burnt releases approximately 400 gm of carbon dioxide, and 1.8 billion cubic metre of wood releases at least 0.6 billion ton of carbon dioxide each year. Wood fuel is, therefore, a potential carbon bomb that may have irreversible impact of the global climate change in the near future.

3.2.4 The Soil Carbon

> 75 billion tons of topsoil with remarkable ability to absorb carbon gets eroded each year much of which goes into the ocean as permanent sink not to be retrieved

Soil was listed as one of the three guarantors of economic wealth alongside labour and capital by famous economist Adam Smith and backed by Thomas Malthus who visualized human catastrophe in the event of exponential population growth. He has been proved wrong so far due to technological interventions, use of artificial fertilizer, soil-less farm culture, genetically modified high output crops, sky-scrappers, fast transportation, etc. Shockingly, the issue of land and soil is back, especially the scarcity of land and soil at specific places making its cost disproportionately high in some places, food production. Farmers with limited cash reserves and no loan facility have sold off highly fertile land to builders for developing cities and townships. The net result is that on the one hand the agriculture land value as % of GDP in countries like USA, Britain, France and Germany has been brought to nought from several hundred per cents in eighteenth century to the present, while the residential property value has risen exponentially between the same periods. Poor land management, expanding cities, exponential population growth and climate change are making soils more prone to erosion and declining rate of soil formation. Since 1960, one-third of global arable land has been lost to erosion. Changing climate may benefit some plants by lengthening growing season in some regions, but the impacts of recurring droughts, floods and pests may reduce the productivity as well. For example while corn production is predicted to be down by 24% by 2050 as compared with 2000 due to reduced production in many areas, but many new areas will switch over to raising corn instead of current crop. Similarly, wheat, rice and potato production will decline by 3, 11 and 9%, respectively, between 2000 and 2050 (Verge et al. 2007).

Large quantities of carbon are stored in living and dead vegetation and soil organic matter, and liberation of this carbon into the atmosphere as carbon dioxide or methane would have a serious impact on global climate. In the soil, carbon exists in two forms, viz., the organic carbon (SOC) and the inorganic carbon (SIC). While SIC is normally found in soil minerals such as carbonates (e.g. limestone or calcium carbonate) commonly found in calcareous or alkali soils, SOC is intricately associated with soil organic matter irrespective of its origin and state of decomposition. This includes plant and animal materials living or dead, in various stages of decomposition, decomposed and burnt. This soil organic matter includes all organic matters of biological origin in a living or decomposed or burnt condition. It includes bacteria, fungi, protozoa, nematodes, mites, termites, ants and earthworms. Soil organic matter typically contains around 60% organic carbon, especially in pristine and old growth forests, deep penetrating roots with slower decomposition rate that facilitate carbon sequestration and improving SOC in deeper layers of soil.

Organic carbon in soil is in a dynamic mode of carbon cycle where the carbohydrates synthesized by plant leaves and stored in root and shoot provide valuable food supply for the growth and reproduction of soil organisms. The organic carbon is eventually converted into humus, and then a part of humus is mineralized to CO_2 and released to the atmosphere. Being a part of the global carbon cycle, the reservoir of SIC[7] in soil is also a dynamic phenomenon depending, inter alia, on the rate of photosynthesis, biological growth of living organisms, decomposition, humus formation and mineralization. Symbiotic mycorrhizal fungi play an important role in building and maintaining SOC by forming association with plant roots, siphoning carbon via root system and improving the absorptive capacity through its fungal hyphae. Most perennial grasses are excellent host for mycorrhizal fungi that add to the process of humus formation in soil. It is estimated that mycorrhizal fungi can store between 4 and 20% of plants total carbon (Bücking et al. 2012). The absence of green cover

---

[7] One ton of soil carbon = 3.67 tons of carbon dioxide emitted or sequestered (Molecular wt of carbon is 12 and that of carbon dioxide is 44. Therefore, carbon dioxide/C = 44/12 = 3.67).

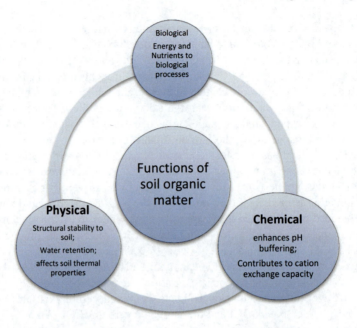

Fig. 3.24 Functions of soil organic matter

will change the carbon dynamics where soil biota will convert most of the SOC and release CO_2 much to the discomfort of humans. Not only that, any decline in soil carbon will negatively impact the physical, chemical and biological function of SOC as described in Fig. 3.24.

Soils are the largest sink of global terrestrial carbon cycle with a capacity to hold 1500 gigatons of carbon (Liddicoat et al. 2010). Unfortunately, large-scale deforestation, conversion to agriculture and use of pesticides have drastically curtailed the soil health. It is, therefore, important to rebuild the SOC through plantations with greater root:shoot ratio, slower decomposing and deep penetrating roots, long rotation crops and use of organic manure. Accumulation of soil carbon has been found to be maximum in deciduous forests with hardwood species or nitrogen-fixing species in tropical or subtropical regions. Long-term forest management regimes such as stocking, weed control, thinning, application of artificial fertilizer, monoculture and fire management also influence accumulation of soil carbon.

Certain plants like maize, cane sugar, sorghum and various tropical grasses are highly efficient in photosynthesis as they are able to bind more CO_2 per unit time, and therefore, these plants are able to utilize most of the internally released CO_2 (formed during photorespiration). Such plants are known as C4 plants as compared to the normal plants which are called C3 plants. C4 plants are mostly found among angiosperms and not in the lower groups of plants such as gymnosperms, ferns, algae, mosses, etc. Most if not all C4 plants are active nitrogen fixers and transport more carbohydrate to the roots and exude more carbon in the rhizosphere. A higher rate of photosynthesis and higher rate of dry matter production by these C4 plants improve the carbon sequestration potential.

3.2.5 Man-made Carbon Source

Cement manufacture is a highly energy-intensive activity, and the energy use by the cement industry is estimated at about 2% of the global primary energy consumption or almost 5% of the global industrial energy consumption. Because of the dominant use of carbon-intensive fuels (primarily coal) in making clinker, the cement industry has become a major source of CO_2 emissions. Besides consuming energy, the clinker-making process also emits CO_2 from the calcination process (Fig. 3.25). In all, the cement industry emits carbon dioxide in three phases

1. Combustion of fossil fuels and calcining the limestone;
2. Consumption of electricity (assuming that the electricity is generated from fossil fuels); and
3. Mobile equipment used for mining of raw material, used for transport of raw material and cement and used at the site.

The data on total carbon dioxide release from the aforementioned three phases is difficult to obtain from the industry and, therefore, is mostly restricted to calcification process.

The reconstruction process started by many countries after the World War II shot up the demand for cement. Between 1970 and 1995, the global production (Fig. 3.26) of cement shot up by almost 850 million tons (from 594 million tons in 1970 to 1453 million tons in 1995) (Worrell et al. 2001). This was mainly (i.e. 83% of global production) confined to 27 countries in North America, Latin America, Europe, Soviet Union, Asia, OECD and Middle East. The carbon emissions from these countries accounted for 5.0% of 1994 world carbon emissions with China in the lead (33.0%), followed by the USA (6.2%), India (5.1%), Japan (5.1%) and Korea (3.7%).

Figures 3.27 and 3.28 provide the details of cement production and CO_2 emissions by Kyoto Annex and Kyoto Non-Annex member nations for the period from 1990 to 2009.

Fig. 3.25 Carbon dioxide release during calcination process

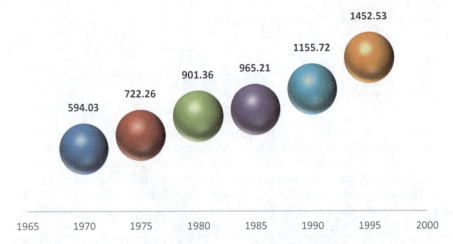

Fig. 3.26 Cement production between 1970 and 1995 (in million tons) (*Source* Worrell et al. 2001)

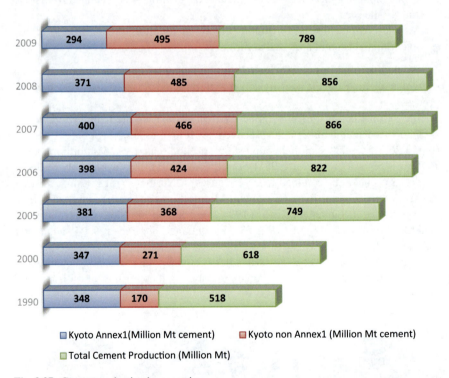

Fig. 3.27 Cement production by countries

3.2 The Carbon Factor

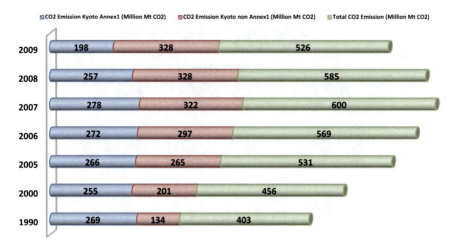

Fig. 3.28 Global carbon dioxide emission by cement companies

3.2.6 Carbon Dioxide in the Ocean

Oceans affect the atmospheric carbon dioxide concentration by the following three mechanisms:

a. **The Solubility Pump**—Absorption and release of carbon dioxide due to changes in solubility of gaseous carbon dioxide;
b. **The Organic Carbon Pump or Biological Pump**—Conversion of inorganic carbon to organic carbon by photosynthesis and export of this organic carbon (called POC—Particulate Organic Carbon) through sinking of organic particles; and
c. **The Calcium Carbonate Counter Pump**—Release of carbon dioxide in surface waters during formation of calcium carbonate shell material by marine organisms.

Carbon dioxide fixation in aquatic ecosystems, particularly oceans, depends on temperature, circulation, salinity and surface stratification. The increase in atmospheric carbon dioxide increases its dissolution in the ocean, and changes in temperature and salinity in ocean affect the chemical equilibrium of gases. Carbon dioxide received by ocean waters from atmospheric gaseous exchange (at ocean–atmosphere confluence) as well as those emanating from animal metabolism and volcanic vents in the seafloor forms carbonic acid. Carbonic acid initially dissociates into bicarbonate and hydrogen ion, and the bicarbonate further releases carbonate and hydrogen ion (Kolbert 2011). Excess hydrogen ions reduce the pH of ocean water, thereby making it acidic. At the same time, excess hydrogen ions combine with carbonate to form bicarbonate and thus reduce carbonate concentration. Loss of carbonate ions affects the formation of calcium carbonate with dramatic consequences for biogenic calcification process. This chemical process essentially indicates that the total dissolved inorganic carbon (DIC) in ocean exists in three major forms:

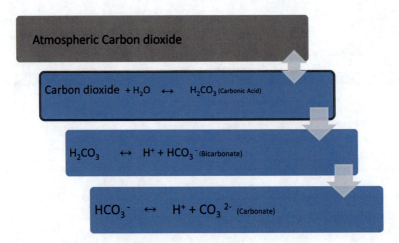

Fig. 3.29 Atmospheric carbon dioxide

- Carbon dioxide
- HCO_3^-
- CO_3^{2-}

The bicarbonate (HCO_3^-) and carbonate (CO_3^{2-}) ions are collectively called dissolved inorganic carbon (DIC). In winters, cold waters at high latitudes enriched with DIC sink to the ocean depths and slowly diffuse upwards into warm surface waters (Fig. 3.29).

The mean pH of ocean surface waters ranges between 7.9 and 8.3 in the open ocean, and a decrease in ocean pH by 0.1 unit corresponds to a 30% increase in the concentration of H^+ ion in sea water (The Fourth Assessment Report of the Intergovernmental Panel on Climate Change (IPCC 2007)).

Under normal circumstances (pre-industrial phase) when atmospheric carbon dioxide was 280 ppm and surface sea water pH was 8.15, the relative proportion of these carbon forms was 0.5% (carbon dioxide), 85% (bicarbonate) and 14% (carbonate) (Seibel and Fabry 2003) (Fig. 3.30).

> **Box 3.2—Startling Facts**
> Earth is spewing 37 billion metric tons of carbon dioxide into the atmosphere every year. How much is actually absorbed and how much remains have not been ascertained with accuracy. However, a spacecraft launched by NASA in July 2014 called Orbiting Carbon Observatory-2 or OCO-2 for short has shown a large plume of carbon dioxide emanating from Northern Australia, Southern Africa and eastern Brazil where forests are being burned for agriculture.
> *The heat-absorbing capacity of ocean is 1000 times greater than that of atmosphere, most of this heat is stored in the upper layers of the ocean, and*

3.2 The Carbon Factor

> *transport of heat by ocean currents has an important bearing on regional climates and thermohaline circulation. Any large-scale changes in the ocean biogeochemistry affect the climate equilibrium of the Earth.*
>
> *Aquatic animals produce carbon dioxide as a metabolic waste which is actively transported to the respiratory organs (gills) and expelled to avoid any reduction in cellular pH. Any imbalance in the level of ocean carbon dioxide will change the gas exchange equilibrium with detrimental effects on the animal.*
>
> *Threats from pollutants gases were known since 1970, but in the absence of scientific certainty and data, the actions were postponed by decades. These pollutants were CO_2, O_3, acid rain, chemicals, petroleum and radioactive material.*
>
> Source: Miller 2015

Fifteen years of analytical research work involving 77,000 sea water samples from different depths around the world has shown that the oceans have absorbed 30% of carbon dioxide released by humans over the past two centuries (Kolbert 2011). The rate of absorption is nearly a million tons per hour. If the process of carbon dioxide release (@30 billion tons/year) continues, the ocean pH will drop from around 8.2 currently to 7.8 by 2100 (Doney et al. 2009). Even if the excess carbon dioxide release is reversed today, it will take a few thousand years to neutralize the changed ocean chemistry. Alternatively, one will need to mine two tons of lime to neutralize one ton of carbon dioxide artificially. Where will this come from (for 30 billion tons of carbon dioxide)? While this may be seen by many nations as a natural phenomenon to counter the effects of excess carbon dioxide released by man, in reality the ocean acidification is an irreversible process with myriad effects including altering the availability of

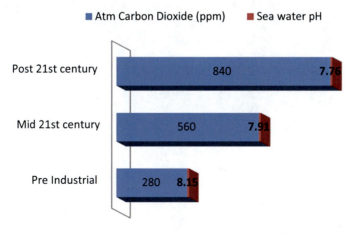

Fig. 3.30 Atmospheric CO_2 and sea water pH (*Source Data* Seibel and Fabry 2003)

key nutrients such as iron and nitrogen, interfering with reproduction in some species and affecting the ability of calcifiers to form shells of calcium carbonate.

In addition to forming carbonic acid, atmospheric carbon dioxide is also consumed by phytoplankton during photosynthesis and a large part of this dissolved organic carbon (DOC) sinks to the bottom of the ocean, a phenomenon called biological pump. Much of the DOC is respired through the action of bacteria and is recirculated to the surface as DIC, while the balance reaches deep-sea ocean sediments. Dissolved organic carbon is produced on daily basis as a part of primary and secondary production in the surface ocean. The process of fixation of inorganic carbon to organic carbon, physical mixing, transport and gravitational export of organic matter from the surface ocean into deeper waters, referred to as '**biological pump**', is the central process in the ocean carbon cycle and requires a higher concentration at the surface than in deeper waters. DOC is produced primarily in the equatorial and coastal upwelling regions, and a combination of factors including organic carbon accumulation at ocean surface, redistribution with the wind-driven circulation, thermohaline circulation at high latitudes and subduction in subtropical gyres eventually leads to transport of DOC to deeper layers.

> **Box 3.3—The Biological Pump**
> Inorganic nutrients and carbon dioxide are fixed during photosynthesis by phytoplankton which in turn release dissolved organic matter. Phytoplankton are consumed by zooplanktons (copepods, amphipods, etc.) which in turn play a significant role by feeding in surface waters and producing faecal pellets. The species composition and mass of zooplankton determine the constituent and sedimentation rate of faecal pellets (organic matter) to the ocean floor. The zooplanktons also transport ingested food in their guts to deep waters where it is further metabolized. Besides, the mortality of zooplankton substantially contributes to the transportation of assimilated organic biomass into deep waters. Simultaneously, DOM is partially consumed by bacteria and respired and the remainder is advected into deep sea as DIC. It is estimated that about 1% of particles leaving the surface ocean reach the seabed and are consumed, respired or buried in the sediments and stored for millions of years. This deep ocean DIC is returned to the atmosphere by thermohaline circulation.
> (Source: Ducklow et al. 2001)

3.2.7 Biological Calcification

Calcium carbonate in the ocean is secreted primarily by planktonic foriminiferans, coccolithophorids, scleractinian corals, pteropod molluscs and calcareous green algae. Other organisms such as bryozoans, echinoderms and calcareous red algae

3.2 The Carbon Factor

secrete calcium carbonate in the form of magnesium calcite. Planktonic foraminiferans and coccolithophorids secrete calcium carbonate in the form of calcite, whereas scleractinian corals, pteropod molluscs and calcareous green algae secrete calcium carbonate in the form of aragonite, a metastable form of $CaCO_3$ (limestone) that is 50% more soluble in sea water than calcite. Coccolithophorids, abundantly found in both coastal water and open ocean, are the major producer of $CaCO_3$. A single bloom in the North Atlantic is reported to cover thousands of square kilometres and produce up to 1 million tons of calcite (Seibel and Fabry 2003).

Marine organisms use carbonate to produce shells of calcite and aragonite (both consisting of calcium carbonate). Currently, the surface ocean is supersaturated with calcite and aragonite (IPCC 2007). Interestingly, the concentration of calcite and aragonite is undersaturated below the 'saturation horizon' that starts at a depth varying between 200 m (in high latitude and Indian Ocean) and 3500 m (in the Atlantic). Calcium carbonate dissolves only when it sinks below the 'saturation horizon' or under the action of biological activity (Table 3.4).

Ocean acidification due to uptake of anthropogenic carbon dioxide may lead to shifts in ecosystem structure and dynamics which may alter the biological production and export from the surface ocean of organic carbon and calcium carbonate to the

Table 3.4 Three main ocean carbon pumps regulate natural carbon dioxide changes by the ocean—the solubility pump, the organic carbon pump and the calcium carbonate counter pump

AIR	SOLUBLE PUMP		ORGANIC CARBON PUMP		$CaCO_3$ Counter Pump	
	carbon dioxide ⇓	carbon dioxide ⇓	carbon dioxide ⇓	carbon dioxide ⇓	carbon dioxide ⇓	carbon dioxide ⇓
SURFACE OCEAN	COLD FRESH High Solubility	WARM HALINE Low	Photosynthesis Carbon Consumption ↑ ↓ Upwelling		Upwelling ↑	Production Of calcareous shells and Alkalinity consumption ↓ Carbon dioxide Flux
INTERMEDIATE WATER						
DEEP OCEAN			Remineralization and Carbon Release		Calcium Carbonate Re-dissolution and Alkalinity Release	
SEDIMENT						

Table 3.5 Fraction of carbon dioxide emissions taken up by the ocean

Time period	Oceanic increase (GtC)[a]	Net carbon dioxide emission
1750–1994	118 ± 19	283 ± 19
1980–2005	53 ± 9	143 ± 10

[a]Sum of emissions from fossil fuel burning, cement production, land use change and the terrestrial biosphere responses
Source IPCC (2007) Table 5.1

ocean floor. Ocean acidification devastates the cup-shaped protective stony exoskeleton of calcium carbonate of tiny coral polyps. Coral polyps are small flower-shaped animals which grow in millions forming a thin layer of living tissue on the surface of the reef. These polyps use calcium ions and carbonate ions for manufacturing skeleton of calcium carbonate. Carbonic acid reacts with carbonate ions making them scarcer in water and thus devoid coral polyps of its raw material. Acidification also affects the regenerative capacity of the coral colonies to produce genetically identical polyps. Many species of coral also engage in 'mass spawning', whereby every polyp releases, in water, a pink sac that contains a sperm and an egg. A lower pH leads to decline in fertilization, in larval development and in attachment of larvae to substratum for producing new colonies. Coral polyps also house symbiotic algae in their cells. These algae use sunlight to synthesize food for polyps which in turn secrete calcium carbonate or limestone. Corals colonies grow best in shallow and clear water with plenty of sunshine to support photosynthesis. Acidification of the ocean makes these corals vulnerable to corrosion, especially the fast-growing branching corals and calcium-secreting algae that help bind the reef.

Failure to form new reefs also means loss of incredible biodiversity of millions of marine species including clams, cucumbers, squirts, fish, turtle and anemones. Besides corals, there are other marine invertebrates such as foraminiferans and coccolithophores (both single-celled organisms), molluscs, echinoderms and crustaceans that act as calcifiers and secrete calcium carbonate to cement coral reefs together. The impact of ocean acidification will be felt more in deeper waters and polar regions where cold water coral will have to survive calcium carbonate under saturated conditions. Exposure of these coral skeletons to acidic waters will lead to their corrosion and fragmentation eventually weakening the base of the reefs (Table 3.5).

3.3 The Janus Faced—Nitrogen

Nitrogen has a good and a bad side. It is an essential component of all living organisms, responsible for controlling biophysical and biochemical activities being an integral component amino acids and nucleic acids {DNA (deoxyribose nucleic acid) and RNA (ribonucleic acid)}. Nonetheless, increase in nitrogen concentration in atmosphere due to extensive use of fertilizers (nitrates and nitrites) and combustion

3.3 The Janus Faced—Nitrogen

processes (nitrogen oxide) has caused ill effects on environment and biodiversity. Nitrogen in the atmosphere (78% by volume) is **inert (non-reactive)** and cannot be used straightaway by the organisms due to strong triple bond of nitrogen atom. This inert nitrogen is converted into reactive nitrogen by processes such as natural fires, burning of fossil fuels, lightning and nitrogen fixation by blue-green algae and bacteria.

Till the early nineteenth century, the reactive nitrogen was mostly available to humans through the process of biological nitrogen fixation. In 1909, Fritz Haber in Germany developed a high-pressure chemical process of artificial nitrogen which was later scaled up to produce industrial nitrogen by Carl Bosch in 1913. This Haber–Bosch process (of synthesizing ammonia from nitrogen and hydrogen) still remains the largest and most economical way of fixing nitrogen artificially. The Haber process now produces 500 million tons (453 billion kilograms) of nitrogen fertilizer per year, mostly in the form of anhydrous ammonia, ammonium nitrate and urea. World natural gas production (3–5%) is consumed in the Haber process that also requires temperature of around 450 °C and pressures of 300 bar, to produce chemical fertilizer responsible for sustaining more than one-third of the Earth's population but with deleterious environmental consequences.

The reactive nitrogen production rate by anthropogenic means has increased ten times when compared with early twentieth-century production. Modern farming practices have increased the release of NH_3, (ammonia), N_2O (nitrous oxide), NO (nitrogen monoxide) and NO_2 (nitrogen dioxide)[8] into the atmosphere and ammonium (NH_4) ions, nitrite (NO_2) ions and nitrate (NO_3) ions into aquatic ecosystems around the world. Use of fossil fuels by households, industries and vehicles also increases the emission of NO and NO_2 (nitrogen dioxide). With an average lifetime of about 120 years and global warming potential 298 times that of an equal mass of carbon dioxide, **nitrous oxide (N_2O) is next most persistent greenhouse gas that contributes to climate change**. The global atmospheric concentration of nitrous oxide is 18% higher than pre-industrial times and has continued to increase by 0.3% per year since 1980 (Gambhir et al. 2017).

The natural nitrogen cycle in the atmosphere is affected by three key nitrogen-containing gases: the NH_3, N_2O and NO_X [(NO, i.e. nitric oxide + NO_2, i.e. nitrogen dioxide)]. Nitrous oxide is the fourth largest contributor to positive radiative forcing and stays in the atmosphere for a long duration. The atmospheric concentration of N_2O has risen from **270 ppb** (parts per billion) in the pre-industrial phase to **319 ppb** in 2005 (van Groenigen et al. 2010). Nitrogen oxides, on the other hand, have short atmospheric life of a few hours to days and help in forming tropospheric ozone. There atmospheric concentration is difficult to measure due to their ephemeral life. They are also responsible for shortening the atmospheric lifetime of methane, thus acting as negative radiating force. The total global emission of nitrogen oxides has increased from 12 TgN/year from pre-industrial period to between 42 and 47 TgN/year in 2000 (Lamarque et al. 2005).

---

[8]NO_2 and NO gases are collectively known as nitrogen oxides.

In essence, inert nitrogen from the atmosphere is converted into ammonia fertilizers to be used for raising food crops. Excess nitrogenous fertilizer pollutes the soils and water sources, increases eutrophication, suffocates plants and animals to death in lakes and estuaries, contaminates groundwater and adds to global warming.

Processes responsible for nitrogen flow through the biosphere, atmosphere, and geosphere.

3.3.1   Nitrogen Fixation and Uptake

It is a process whereby nitrogen is converted into ammonium through nitrogen-fixing bacteria (e.g. Rhizobium) which form symbiotic relationship with host plants particularly of legume family. These bacteria form root nodules and fix nitrogen present in the soil for use by plants by converting into ammonia which is quickly incorporated into protein and other nitrogen compounds. In addition, there are also nitrogen-fixing bacteria that exist without plant hosts and are known as free-living nitrogen fixers. In aquatic environments, blue-green algae (in reality a bacteria called cyanobacteria) are an important free-living nitrogen fixer. Nitrogen in nature is also fixed by high-energy natural events such as lightning, forest fires and volcanic eruptions. These events break the triple bond of nitrogen molecules and make individual nitrogen atoms available for transformation.

3.3.2   Nitrogen Mineralization

After the death of the plants, certain decomposers (bacteria and fungi) present in the soil consume the organic matter and convert the nitrogen present in them to ammonium (NH_4^+). Once in the form of ammonium, nitrogen is available for use by plants or for further transformation into nitrate (NO_3^-) through the process called nitrification.

3.3.3   Nitrification

Ammonium (NH_4^+) is oxidized to nitrate in two stages by aerobic chemo-autotrophic bacteria called Nitrosomonas and Nitrobacter (Fig. 3.31).

These three forms of nitrogen (ammonium, nitrite and nitrate) are the most common **reactive forms** of dissolved inorganic nitrogen in aquatic ecosystems. Since oxygen is required for nitrification, it can happen only in oxygen-rich environments like circulating or flowing waters and the surface layers of soils and sediments. The process of nitrification has some important consequences. Ammonium ions are positively charged and, therefore, stick to negatively charged clay particles and soil

Fig. 3.31 Sources of greenhouse gas emission

organic matter. The positive charge prevents ammonium ion from being washed out of the soil (or leached) by rainfall. In contrast, the negatively charged nitrate ion is not held by soil particles and is washed down the soil profile, leading to decreased soil fertility and nitrate enrichment of downstream surface and groundwater. Reactive nitrogen (like NO_3^- and NH_4^+) present in surface waters and soils can also enter the atmosphere as the smog component nitric oxide (NO) and the greenhouse gas nitrous oxide (N_2O).

3.3.4 Denitrification

Oxidized forms of nitrogen, i.e. nitrite and nitrate, are converted to dinitrogen (N_2) and nitrous oxide (N_2O) through an anaerobic process that is carried out by denitrifying bacteria (Achromobacter, Bacillus, Micrococcus, Pseudomonas) (Fig. 3.32).

Nitric oxide (NO) and nitrous oxide (N_2O) are both environmentally sensitive in the sense that nitric oxide contributes to smog and nitrous oxide is an important greenhouse gas. Once N_2 (dinitrogen) gas is formed from nitrite and nitrate, it is quickly lost to the atmosphere and thus remains unavailable for conversion to biological form. Denitrification is the only nitrogen transformation mechanism that removes nitrogen from ecosystems (almost irreversibly), and it roughly balances the amount of nitrogen fixed by the nitrogen fixers as described above.

Nitrogen in the atmosphere can be blown into terrestrial environments, causing long-term changes. For example, nitrogen oxides in the acid rains have caused forest death and decline in parts of Europe and the Northeast United States. Increases in atmospheric nitrogen can bring in imperceptible changes in dominant species and ecosystem function in some forest and grassland ecosystems. For example, on nitrogen-poor serpentine soils of Northern Californian grasslands, plant assemblages have historically been limited to native species that can survive without a lot of nitrogen. There is now some evidence that elevated levels of atmospheric nitrogen

Fig. 3.32 Composition of atmospheric gases (%)

input from nearby industrial and agricultural development have paved the way for invasion by non-native plants. As noted earlier, NO is also a major factor in the formation of smog which is known to cause respiratory illnesses like asthma in both children and adults.

3.3.5 Major Anthropogenic Sources of Inorganic Nitrogen in Aquatic Ecosystems

Point Sources:

- Wastewaters from livestock (cattle, pigs, chickens) farming;
- Nitrogen releases from aquaculture (fish, prawns, shrimps) operations;
- Municipal sewage effluents (including effluents from sewage treatment plants that are not performing tertiary treatments);
- Industrial wastewater effluents;
- Run-off and infiltration from waste disposal sites;
- Run-off from operational mines, oil fields and unsewered industrial sites; and
- Overflows of combined storm and sanitary sewers.

Non-point Sources:

- Widespread cultivation of nitrogen-fixing crop species and the subsequent nitrogen mobilization among terrestrial, aquatic and atmospheric realms;
- Use of animal manure and inorganic N fertilizers, and the subsequent run-off from agriculture;
- Run-off from burned forests and grasslands;
- Run-off from nitrogen-saturated forests and grasslands;
- Urban run-off from unsewered and sewered areas;
- Septic leachate and run-off from failed septic systems;
- Run-off from construction sites and abandoned mines;
- Nitrogen loadings to groundwater and, subsequently, to receiving surface waters (rivers, lakes, estuaries, coastal zones); and
- Emissions to the atmosphere of reduced (from volatilization of manure and fertilizers) and oxidized (from combustion of fossil fuels) N compounds, and the subsequent atmospheric (wet and dry) deposition over surface waters.
Source: Camargo and Alonso (2006)

Other activities that can mobilize nitrogen (from long-term storage pools) include biomass burning, land clearing and conversion, and wetland drainage.

In the intensively managed agriculture farms, nitrogen fertilizer is overused by 30–60% in the expectation of improved productivity. Excess nitrogen fertilizer invariably gets washed off of agriculture fields and accumulates in surface and groundwater. In surface waters, this extra nitrogen can lead to nutrient overenrichment, particularly in coastal waters receiving the inflow from polluted rivers. A recent survey of 40 lakes in China showed that more than half of these had excess nitrogen or phosphorus

3.3 The Janus Faced—Nitrogen

(Charles 2013). High concentration of reactive nitrogen in the form of ammonium ions (NH_4^+), nitrite ions (NO_2^-) and nitrate (NO_3^-) derived from human activities enhances the proliferation of primary producers (phytoplankton, macrophytes and benthic algae) in the aquatic ecosystems by creating dead zones in which algae and phytoplankton bloom, die and decompose using up oxygen and suffocating fish and other aquatic fauna.

> **Box 3.4—Health Effects**
> The addition of nitrogen bond in the environment can change the composition of species due to susceptibility of certain organisms to the consequences of nitrogen compounds, and food rich in nitrogen compounds can reduce oxygen transport of the blood. Nitrates in animal stomach can form nitramines, a dangerous carcinogenic compound. Nitrates and nitrites from polluted drinking water may induce methemoglobinemia in humans, a condition that blocks the oxygen-carrying capacity of haemoglobin, resulting in the formation of methemoglobin.

Global ammonia capacity was 153 m tons/year NH_3 in 2009, with the main additions occurring in China, Trinidad, Indonesia, Oman, India and Egypt. IFA estimated global nitrogen supply capability (or effective capacity) to grow from 134.8 m tons N in 2010 to 158.5 m tons N in 2014. India is ranked as the world's third largest producer and importers of ammonia. Imports which are estimated at 1.9 m tons in

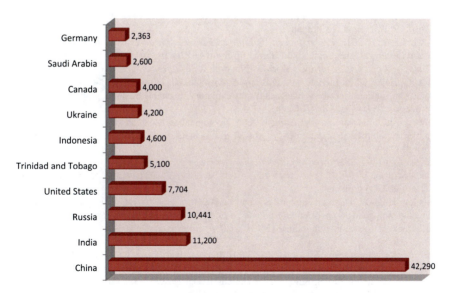

Fig. 3.33 Top ten ammonia-producing countries in 2009 (thousand metric ton) (*Source* Marland 2010)

2009 are used for phosphate fertilizers since domestic ammonia production is mostly integrated with urea production. Excluding any plant restarts, IFA expected India's ammonia capacity to reach 20 m ton/year in 2014. Trinidad is the world's largest ammonia exporter with a capability close to 4.9 m tons in 2009. Other Caribbean and Latin American exporters include Venezuela, Mexico, Argentina, Brazil and Colombia (Fig. 3.33).

> Take care of your earth,
> Look after its creatures.
> Don't leave your children,
> A planet that's dead.
> —Lakshmi Shree, India, age 12

References

Abdel-Motalib, Nageh Khalaf Allam. 2009. *An Investigation into the Doping and Crystallinity of Anodically Fabricated Titania Nanotube Arrays: Towards an Efficient Material for Solar Energy Applications*. The Pennsylvania State University.

Abramovitz, Janet N. 2001. Averting unnatural disasters. *State of the World* 123–142.

Agarwal, Anil, Sunita Narain, and Srabani Sen (eds.). 1999. *The State of India's Environment: The Citizens' Fifth Report. Part I: National Overview. Part II: Statistical Database*. Centre for Science and Environment pp. 182–183.

Blunden, J., D. S. Arndt, and G. Hartfield (eds.). 2018. State of the climate in 2017. *Bulletin of the American Meteorological Society* 99(8): Si–S332. https://doi.org/10.1175/2018bamsstateoftheclimate.1.

Brown, Lester R. 2008. *Plan B 3.0: Mobilizing to Save Civilization (substantially revised)*. New York: W. W. Norton & Company.

Brown, Lester R. 2013. *Eco-economy: Building an Economy for the Earth*. Abingdon: Routledge.

Brown, Lester R., Janet Larsen, and Bernie Fischlowitz-Roberts. 2013. *The Earth Policy Reader: Today's Decisions, Tomorrow's World*. Abingdon: Routledge.

Bücking, Heike, Elliot Liepold, and Prashant Ambilwade. 2012. The role of the mycorrhizal symbiosis in nutrient uptake of plants and the regulatory mechanisms underlying these transport processes. In *Plant science* (InTech).

Camargo, Julio A., and Álvaro Alonso. 2006. Ecological and toxicological effects of inorganic nitrogen pollution in aquatic ecosystems: A global assessment. *Environment International* 32 (6): 831–849.

Charles, Dan. 2013. Fertilized world. *National Geographic*.

Doney, Scott C., Victoria J. Fabry, Richard A. Feely, and Joan A. Kleypas. 2009. Ocean acidification: The other CO_2 problem.

Ducklow, Hugh W., Deborah K. Steinberg, and Ken O. Buesseler. 2001. Upper ocean carbon export and the biological pump. *Oceanography-Washington D.C.-Oceanography Society* 14 (4): 50–58.

Earth System Research Laboratory (ESRL). 2016. ftp://ftp.cmdl.noaa.gov/ccg/co2/trends/co2_mm_mlo.txt

Energy Information Administration (EIA). 2007. International Energy Outlook 2007.

Energy Information Administration (EIA). 2010. Annual Energy Outlook 2010.

Food and Agriculture Organization of the United Nations (FAO). 2006. Global Forest Resources Assessment 2005. FAO Rome.

References

Gambhir, Ajay, Tamaryn Napp, Adam Hawkes, Lena Höglund-Isaksson, Wilfried Winiwarter, Pallav Purohit, Fabian Wagner, Dan Bernie, and Jason Lowe. 2017. The contribution of non-CO_2 greenhouse gas mitigation to achieving long-term temperature goals. *Energies* 10 (5): 602.

Hawkins, Ed, and Phil Jones. 2013. On increasing global temperatures: 75 years after Callendar. *Quarterly Journal of the Royal Meteorological Society* 139(677): 1961–1963.

Hulme, Mike. 2009. On the origin of 'the greenhouse effect': John Tyndall's 1859 interrogation of nature. *Weather* 64 (5): 121–123.

IEA GHG. 2002. *Transmission of CO_2 and Energy*. Cheltenham, International Energy Agency Greenhouse Gas R&D Programme. Cheltenham.

International Energy Agency (IEA). 2006. World energy outlook 2006. Organisation for Economic Co-operation and Development (OECD)/International Energy Agency (IEA).

International Energy Agency (IEA). 2010. World energy outlook 2010. Organisation for Economic Co-operation and Development (OECD)/International Energy Agency (IEA).

IPCC. 2007. IPCC Fourth Assessment Report: Climate Change 2007 Synthesis Report. Valencia.

Kolbert, Elizabeth. 2011. The acid sea. *National Geographic Magazine* 100–131.

Koroneos, Christopher, and Agis M. Papadopoulos. 2010. Systems to support decisions on electric power generation. *Environmental Systems-Volume III* 141.

Lamarque, J-F., J. T. Kiehl, G. P. Brasseur, T. Butler, P. Cameron-Smith, W. D. Collins, and W. J. Collins et al. 2005. Assessing future nitrogen deposition and carbon cycle feedback using a multi-model approach: Analysis of nitrogen deposition. *Journal of Geophysical Research: Atmospheres* 110(D19)

Liddicoat, Craig, Schapel, Amanda, Davenport, David and Dwyer, Elliot. June 2010. PIRSA discussion paper – Soil carbon and climate change.

Ma, Jian. 2011. Techno-economic analysis and engineering design consideration of algal biofuel in southern Nevada.

Marland, G. Boden, T. A. and Andres, R. J. 2013. Global, Regional, and National CO_2 Emissions, Trends: A Compendium of Data on Global Change (Oak Ridge, TN: Carbon Dioxide Information Analysis Center, 2010); 2008 and 2009 emissions calculated by Earth Policy Institute from energy consumption in BP, Statistical Review of World Energy (London: 2010), and cement production in U.S. Geological Survey (USGS), Mineral Commodity Summaries 2010 (Washington, D.C.: 2010), p. 39; USGS, Mineral Commodity Summaries 2009 (Washington, D.C.: 2009), p. 41.

Mohammed, Jamal, and Anthony Kofi Osei-Fosu. 2016. Fuel wood commercialization and households welfare in the northern region of Ghana: An implication for rural livelihood improvement. *International Journal of Energy and Environmental Research* 4(3): 54–79.

Nelsen, Matthew P., William A. DiMichele, Shanan E. Peters, and C.Kevin Boyce. 2016. Delayed fungal evolution did not cause the Paleozoic peak in coal production. *Proceedings of the National Academy of Sciences* 113 (9): 2442–2447.

Pachauri, Rajendra K., and Andy Reisinger. 2007. IPCC fourth assessment report. IPCC, Geneva (2007).

Prichard, Bob. 2010. The AMS Weather Book: The Ultimate Guide to America's Weather. By J. Williams Published by Chicago University Press, 2009 Hardback 368 pp ISBN 978-0226898988. *Weather* 65(4): 97–97.

Raupach, Michael R., Gregg Marland, Philippe Ciais, Corinne Le Quéré, Josep G. Canadell, Gernot Klepper, and Christopher B. Field. 2007. Global and regional drivers of accelerating CO_2 emissions. *Proceedings of the National Academy of Sciences* 104(24): 10,288–10,293.

Reddy, P.Jayarama. 2013. *Clean Coal Technologies for Power Generation*. Boca Raton, FL: CRC Press.

Rubin E, De Coninck H. 2004. *IPCC special report on carbon dioxide capture and storage*. UK: Cambridge University Press. TNO (2004): Cost Curves for CO_2 Storage, Part 2. 14 Nov 2005.

Schneider, Stephen H. 1989. The greenhouse effect: Science and policy. *Science* 243 (4892): 771–781.

Schneider, Thomas W. 2004. *Expanded Programme of Work on Forest Biological Diversity Versus IPF/IFF Proposals for Action*. No. 2004/1. Work Report of the Institute for World Forestry.

Schoennagel, Tania, Jennifer K. Balch, Hannah Brenkert-Smith, Philip E. Dennison, Brian J. Harvey, Meg A. Krawchuk, Nathan Mietkiewicz, et al. 2017. Adapt to more wildfire in western North American forests as climate changes. *Proceedings of the National Academy of Sciences* 114 (18): 4582–4590.

Seibel, Brad A., and Victoria J. Fabry. 2003. Marine biotic response to elevated carbon dioxide. *Advances in Applied Biodiversity Science* 4: 59–67.

Sieminski, Adam. (2014). International energy outlook. *Energy Information Administration (EIA)* 18.

Smith KR., Uma R., Kishore VVN., Lata K., Joshi V., Zhang J., Rasmussen RA., Khalil MAK. 2000. *Greenhouse gases from small-scale combustion devices in developing countries, phase IIa: household stoves in India*. EPA-600/R-00-052, U.S. Environmental Protection Agency, Office of Research and Development, Washington, DC.

Solomon, Susan, Dahe Qin, Martin Manning, Zhenlin Chen, Merlinda Marquis, Kristen B. Averyt, M. Tignor, and Henry L. Miller. 2007. Contribution of working group I to the fourth assessment report of the intergovernmental panel on climate change, 2007.

Srivastav, Asheem, and Suvira Srivastav. 2015. *Ecological Meltdown: Impact of Unchecked Human Growth on the Earth's Natural Systems*. The Energy and Resources Institute (TERI).

Stone, Daniel. 2014. The Afterlife of a Landfill. *National Geographic* 225 (6): 26.

United Nations Development Programme (UNDP). 2000. World Energy Assessment: Energy and the Challenge of Sustainability.

United Nations Environment Programme (UNEP). 2013. Global environment outlook 2000. Vol. 1. Abingdon: Routledge.

van Groenigen, J.W., G.L. Velthof, O. Oenema, K.J. Van Groenigen, and C. van Kessel. 2010. Towards an agronomic assessment of N_2O emissions: A case study for arable crops. *European Journal of Soil Science* 61 (6): 903–913.

van Lierop, Pieter, Erik Lindquist, Shiroma Sathyapala, and Gianluca Franceschini. 2015. Global forest area disturbance from fire, insect pests, diseases and severe weather events. *Forest Ecology and Management* 352: 78–88.

Verge, X.P.C., C. De Kimpe, and R.L. Desjardins. 2007. Agricultural production, greenhouse gas emissions and mitigation potential. *Agricultural and Forest Meteorology* 142 (2–4): 255–269.

Watson, R. T., L. G. Meira Filho, E. Sanhueza, and A. Janetos. 1992. Greenhouse gases: Sources and sinks. *Climate Change* 92: 25–46.

Watson, R.T., H. Rodhe, H. Oeschger, and U. Siegenthaler. 1990. Greenhouse gases and aerosols. *Climate Change: The IPCC Scientific Assessment* 1: 17.

Weart, Spencer R. 2008. *The Discovery of Global Warming*. Cambridge, MA: Harvard University Press.

World Bank. 2006. Development Economics Dept. Development Data Group, and World Bank. Environment Dept. *The Little Green Data Book...: From the World Development Indicators...* International Bank for Reconstruction and Development/The World Bank.

Worrell, Ernst, Lynn Price, Nathan Martin, Chris Hendriks, and Leticia Ozawa Meida. 2001. Carbon dioxide emissions from the global cement industry. *Annual Review of Energy and the Environment* 26(1): 303–329.

Chapter 4
Natures' Reaction to Anthropogenic Activities

Abstract Climate-related events including catastrophic ones are a natural phenomenon, but all natural events cannot be directly linked to the recent climate change. It is not possible to say with confidence that a particular storm, fire, flood, drought or earthquake was caused by the recent climate change. But what can be said with certainty is the fact that any action that leads to increase in GHG, whether locally or globally, increases the risk of catastrophic events. It is also true that in recent years, there has been rise in torrential hurricanes, droughts, and ice and snow storms, raging heatwaves that have devastated many areas of the world claiming human and animal lives. Disproportionate increase in human population, overconsumption of resources and pollution has reached a point where rate of replenishment of natural resources is far less than depletion. Different regions and countries in the world are under some threat or the other including inundation by sea level rise, melting of snow and glacier retreat, forest fire, droughts. The cumulative economic impact of restoration may run into billions of dollars which many poor nations can ill afford.

Keywords Catastrophe · Ecological footprint · Extreme weather events
Hurricane · Drought · Forest fire · El Nino · La Nina · Bush meat
Medicinal plant · Water distress

4.1 Increased Vulnerability to Natural/Man-Made Disasters

Almost every individual on this planet is conversant with the proverb '*whatsoever a man soweth, that shall he also reap*'. In other words, it means '*you will get what you sow*'. There is a general belief that ocean and atmosphere never keep anything unwanted in their belly and return it soonest. The scientific logic behind this is easy to understand. The total energy of the Earth is well balanced, and any shift in that balance in any of its components results in destabilization inimical to the flow of energy. Greater the destabilization larger the reaction of nature that may be IRREVERSIBLE, no one knows for sure. A substantial proportion of 7 billion plus

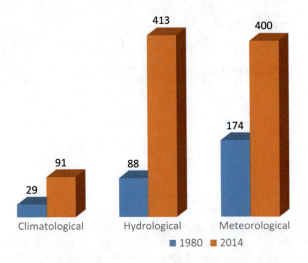

Fig. 4.1 A comparison of catastrophic events

people have forced so many changes in global energetic that it seems difficult to retrieve the situation, at least in the near future.

Climate-related all events including catastrophic ones are a natural phenomenon, and all natural events cannot be directly linked to the recent climate change. It is not possible to say with confidence that a particular storm, fire, flood, drought or earthquake was caused by the recent climate change. But what can be said with certainty is the fact that any action that leads to increase in GHG, whether locally or globally, increases the risk of catastrophic events. It is like saying that there may not be a direct correlation between smoking and lung cancer (not every smoker has lung cancer) but that smoking increases the chances of lung cancer. What is worrisome is the fact that in recent years, there has been rise in torrential hurricanes, droughts, and ice and snow storms, raging heatwaves that have devastated many areas of the world claiming human and animal lives.

The climate events (Fig. 4.1) can be divided into:

1. Climatological events: such as extreme temperatures, droughts and fires. For instance, in May 2015 global average temperature was highest on record. Some 2200 died people in India alone when the temperature soared 45 °C during the heatwave of 10 days.
2. Hydrological events: such as floods and mudslides;
3. Meteorological events: Tropical, extra-tropical and local storms. For example, cyclone Marian killed 140,000 people in 1991 and Cyclone Sidr destroyed 565,000 homes. A nine-feet storm surged in 2012 from Hurricane Sandy hit New York City where water level went up to 14 ft. higher than normal.

The ongoing climate change is a reactive process, the degree of reaction varying with the quantum of societal actions. The societies have to necessarily take corrective actions for survival and development. Or else they perish. Since climate change is a slow process that commences with freak weather pattern, societies normally take long time to understand and appreciate the impact and therefore corrective measures are delayed sometimes reaching the point of no return. The Akkadian empire established around 4300 BC between Tigris and Euphrates rivers suddenly collapsed after a century for reasons that remained a mystery for thousands of years (Kolbert 2009). Recently, the climate scientist, while examining past records, came to the conclusion that the Akkad's collapse was caused by a devastating drought and shift in rainfall which also led to the fall of Old Kingdom of Egypt. In almost the same manner, the Tiwanaku civilization near Lake Titicaca in the Andes and famous Mayan civilization disintegrated around AD 1100 and AD 800, respectively (Binford et al. 1997). Around 2200 BC, a shift in the Mediterranean westerly winds and a reduction in the Indian monsoon for 300 years adversely affected agriculture from Aegean Sea to Indus River halting the development in the cities along the northern fringes of Euphrates and converting them into ruins as the inhabitants deserted. The probable cause of demise of all these civilizations is believed to be drastic changes in climate triggered by natural factors hitherto unknown to modern science.

Majority of the seven billion people occupying this spaceship have accepted the reality of human-induced (Fig. 4.2) climate change and are aware of the fact that if left unmanaged, this change will reverse the growth, development and well-being of the current as well as future generations. Impact of varying degrees is being frequently felt in all parts of the close-knit world. People are becoming aware that sea level rise will inundate populations and infrastructure along the coast from Florida to Dhaka; that erratic rainfall and severe droughts in semi-arid Africa will aggravate food insecurity and malnourishment; that fast-receding Himalayan and Andean glaciers which regulate water flow provide drinking water and generate hydropower will devastate rural and urban poor; that disappearance of natural forests will increase soil loss; that destruction of mangroves will expose the coast and make them vulnerable to cyclones.

Our ever-increasing ecological footprints continue to impede the natural recovery processes in ecosystems which in turn affects numerous sectors and productive environment in agriculture, forestry, energy and coastal zones in developing and developed countries. No one is immune from the impact as adverse climate trends, variability and shocks do not discriminate by income. More than 70,000 people perished in rich Europe during the summer heat of 2003 (Stone et al. 2010), and people in sub-Saharan Africa, South and East Asia witnessed increase in malaria and dengue in the past decade. Aral Sea almost disappeared due to overconsumption of water for irrigating cotton crop in water-stressed areas of Central Asia. Mangrove clearance in many countries for shrimp farming and development exposed their coast as was evident during the December 2004 Tsunami. Hurricane Katrina completely exposed the US preparedness, planning and adaptation. 1970 tropical storm in Bangladesh decimated 500,000 people.

The Cumberland River rose by 52 ft. in 48 h during May 2010 as a result of extraordinary torrential rains as a result of sudden build-up of warm and humid air from the Gulf of Mexico. Loss of nearly two billion dollars apart, the city was thrown out of gear for several months with terrified residents possibly anticipating another such event sooner rather than later. Similarly, in April 2010, the city of

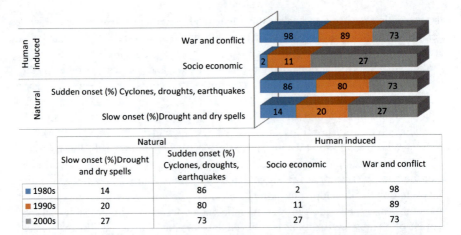

Fig. 4.2 Natural and human-induced disasters (*Source* Adapted from The State of Food insecurity in the World 2008. FAO 2008)

Rio de Janeiro suffered heavy losses due to 11 in. downpour in 24 h triggering mudslides and taking heavy toll on lives and property. A total of 20 million people were affected in Pakistan during the flood of July–August 2010. Every year such weather-related extreme events are affecting some part of the globe with varying intensity and frequency. Under normal conditions, a large part of the Sun's energy is used to evaporate water from water bodies, plants and soil. But when water bodies dry up, natural vegetation is destroyed and soil is parched, and the Sun's energy heats up air and water vapour in the atmosphere further pumping heat into atmosphere and big storms such as hurricanes and tornadoes. A continued warming trend is expected to increase the strength of tornadoes and hurricanes by 2–11% by 2100 (Miller 2012).

Every nation impacted by extreme weather events suffers losses (Table 4.1) that depend upon factors such as population density, infrastructure, resilience or vulnerability and preparedness. Losses due to floods, cyclones and other natural disasters during 2014 in Asia itself were nearly 50 Billion USD in 2014 (The Economist 2015c). An assessment of natural disasters in the USA indicates economic losses to the extent of 339 billion USD in 46 catastrophes between 1980 and 1995. A similar assessment of fifteen years between 1996 and 2011 shows a cumulative loss of USD 541 billion from 87 disasters. These statistics clearly indicate that the number of catastrophic events almost doubled during late twentieth and early twenty-first century. Recent data of 2017 shows that there were 16 weather and climate disaster events with losses exceeding $1 billion each across the USA. Overall, these events resulted in the deaths of 362 people and had significant economic effects on the areas impacted (Smith and Richard 2013). Not many countries meticulously maintain such statistics. But the general trend clearly indicates increase in extreme weather events with massive economic, human and other losses (Fig. 4.3).

4.1 Increased Vulnerability to Natural/Man-Made Disasters

Table 4.1 Major economic losses in the USA due to extreme weather events

Year	Event	Billion USD
1980	Drought/heatwave	56
1988–89	Drought/heatwave	78
1993	Hurricane (Andrew)	44
1994	Floods	33
2004	Hurricane (Charley)	18
2004	Hurricane (Ivan)	17
2005	Hurricane (Katrina)	146
2005	Hurricane (Rita)	19
2005	Hurricane (Wilma)	19
2008	Hurricane (Ike)	29

(*Source* Miller 2012)

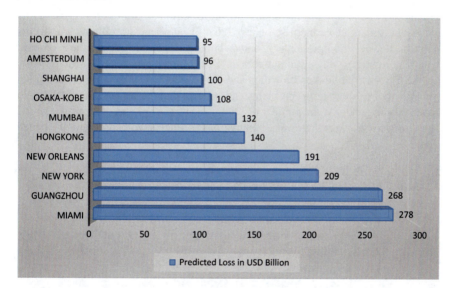

Fig. 4.3 Economic loss in 2050 if the extreme weather events continue (*Source* Hallegatte et al. 2013)

While rich population may recover quickly from losses as their life and assets are insured, it is the poor who will be hard hit for two reasons: one—direct loss to life and property and two—indirect loss in income that flows from rich to poor.

4.2 Population and Extreme Weather Nexus

The 1972, UN Conference on the Human Environment in Stockholm proclaimed that

> the natural growth of human population continuously presented problems for the preservation of the environment. Man's capability to transform his surroundings, if used wisely, can bring to all peoples the benefits of development and the opportunity to enhance the quality of life. Wrongfully or heedlessly applied, the same power can do incalculable harm to human beings and the human environment.

Paul Ehrlich, a renowned biologist having wonderful foresight, suggested in his book 'The Population Bomb' (published in 1968) that the world population growth was the basic causal force for pollution and hunger. He cautioned the global community for immediate decisive action against increasing human numbers and advocated family planning, agricultural development, and industrialization in less developed countries to avoid famine and death, as well as the need for greater diffusion of conservationist values towards the natural world. Contrary to Ehrlich's suggestion, the late Julian Simon, an economist, believed that the ecological problems will decline in time as new technologies emerge and technologists learn to expand the economies in more enlightened ways. Julian wrote in 'The Ultimate Resource 2', (Simon 1996) that the problems of hunger, crowding and energy shortages associated with population growth are 'short-term' problems and insisted that the bigger the population of a country, the greater is the number of scientists and the amount of scientific knowledge produced to exert their own imaginations for their own benefit and the benefit of all.

> **Box 4.1 The Problem of Population as explained by Paul Ehrlich**
> (Source: Ehrlich 1970)
> 'I have understood the population explosion intellectually for a long time. I came to understand it emotionally one stinking hot night in Delhi a couple of years ago....The temperature was well over 100, and the air was a haze of dust and smoke. The streets seemed alive with people. People eating, people washing, people sleeping. People visiting, arguing and screaming. People thrusting their hands through the taxi window, begging. People defecating and urinating. People clinging to buses. People herding animals. People, people, people, people'.

Ignoring the views of economists like Julian Simon, the report on the State of World Population issued by the UN Family Planning Association in 1990 appealed to all the populated nations to slow down their population growth in order to protect environment and attack poverty. It stressed the need for family planning, education and awareness and health care coupled with the improved status of women in slowing population growth. The effect of UN appeal can be seen in the demographic data,

4.2 Population and Extreme Weather Nexus

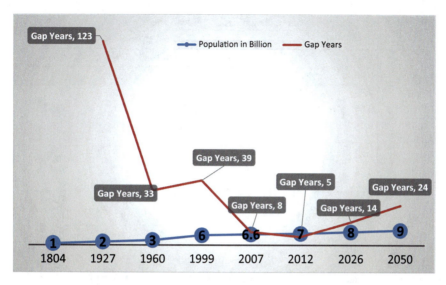

Fig. 4.4 Time lag in human population growth (*Source* Haub and Diana 2007)

and estimates of 2007 which indicate that the global rate of natural[1] increase is 1.2% (*0.1% for more developed and 1.5% for less developed regions*). The estimates forecast that the global population will touch 8 billion by mid-2025 and 9.2 billion by mid-2050 (Fig. 4.4) instead of earlier prediction of 10 billion (Haub and Cornelius 2007). While the population growth rate appears to have slowed down, as is evident from the above assessments, the actual number continues to increase, though at a slower pace than anticipated during mid-seventies and eighties.

Joel Cohen in his book 'How many people can the Earth support' points out that it is almost impossible to estimate the carrying capacity without knowing the technologies, the distribution of material, the political institutions and economic arrangements (Cohen 1995). More recent estimates by David Pimentel suggest a figure of 1 to 2 billion people in relative prosperity while other scholars suggest a range of 4–16 billion (Bradshaw and Brook 2014). Food and Agriculture Organization has, however, estimated that 33 billion people can be fed on minimum rations using every available hectare of suitable land for high-intensity food production. FAO estimate, however, fails to take cognizance of simple realities that modern agriculture uses land to convert petroleum into food thanks to heavy dependence of agriculture on petrochemicals, fertilizers and pesticides. And with millions of people still living in chronic hunger, it is difficult to accept that Earth can feed 33 billion without ecological catastrophe. Though the rate of population growth is slower than anticipated (Table 4.2) by many demographers, nearly 2 billion people are expected to be added to our world during the first half of this century. Ironically, 50% of the future population will be added in

---

[1] Natural increase means the birth rate minus the death rate, implying annual rate of population growth without regard for migration.

Table 4.2 Current and projected population size and growth rates

	Population			Annual growth rate		
	1985	2000	2025	1950–1985	1985–2000	2000–2025
Region	(Billion)			(Per cent)		
World	4.8	6.1	8.2	1.9	1.6	1.2
Africa	0.66	0.87	1.62	2.6	3.1	2.5
Latin America	0.41	0.55	0.78	2.6	2.0	1.4
Asia	2.82	3.55	4.54	2.1	1.6	1.0
North America	0.26	0.30	0.35	1.3	0.8	0.6
Europe	0.49	0.51	0.52	0.7	0.3	0.1
USSR	0.28	0.31	0.37	1.7	0.8	0.6
Oceania	0.02	0.03	0.04	1.9	1.4	0.9

sub-Saharan Africa and 30% in South and Southeast Asia, the regions already hard hit by drought, heat, extreme weather, pollution and poor economic growth.

Global population is increasing by about 70–80 million each year primarily due to high proportion of young population (between the age group 15 and 24 years) in all continents except Europe. Population in this group went up from 769 million in 1985 to 1.05 billion in 2000 and is expected to reach 1.16 billion by 2050. This high-fertility group has important implications for future population growth, migration and consumption pattern. Over the next 20–25 years, three-fourths of population growth in developing countries and most of the population growth in East Asia will be determined by the young age composition that will have propensity for migration, seeking better education and employment opportunities in the megacities.[2] In many such developing nations, rural areas are increasingly becoming clusters of low-productivity marginal lands due to rural–urban migration by younger age group and expansion of urban set-up. Lower levels of agricultural productivity and income exert greater pressures at local scales inducing people to behave like myopic, making use of local resources without consideration for national, regional and global effects or implications for future generations. For instance, high population density and poverty increased the cultivated area in upland forest in the Philippines from 10% in 1957 to 30% in 1987, primarily due to increased demand for agricultural and fuel-wood resources (Hunter 2000). Brazil, with 35% of world's rainforests, has suffered maximum deforestation due to migration for expansion of agriculture and industrial growth following the construction of two highways (BR-364 to Rondonia and the Trans-Amazon highway) (United Nations 2001). On a global scale, nearly 60% of recent deforestation is attributable to agriculture sector, 20% to logging (including mining and petroleum) and 20% to household sector for fuel-wood (United Nations 2001).

---

[2]Megacity—A city with more than 10 million population. The number of megacities increased from 5 in 1975 to 23 in 2005.

4.2 Population and Extreme Weather Nexus

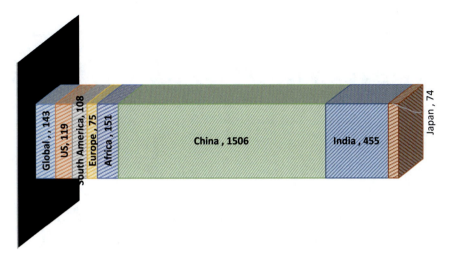

Fig. 4.5 Per cent GDP growth between 1980 and 2009 (*Source* Kunzig 2011)

The rapid population growth will continue to overstrain the Earth's natural resources as everyone wants to be well fed, clothed, housed, and have access to electricity, clean air, clean water and sanitation. To meet these necessities, water, land, forests and other natural resources will be exploited unsustainably. More forest will be cut down to provide wood for housing and fuel, and more cleared land will be needed for agriculture and infrastructure development. At the same time, massive waste from human populations will continue to pollute our ecosystems. Logically, even if the markets function with efficiency with the support of state of the art technology, it will take more resources to support a larger population than a smaller one, and the environmental costs of doing so will be insurmountable.

The global GDP increased from $29.8 trillion in 1980 to $72.5 trillion in 2009 indicating improved standards of living on the one hand and higher consumption of energy and more waste generation on the other (Fig. 4.5). The global economy is predicted to reach $150 trillion by 2030 with countries like China maintaining 8% GDP per annum (Hawksworth and Gordon 2006). This means that on the one hand, rapidly expanding human population in developing world will change the consumption dynamics with greater energy demand for vehicles, air conditioners, houses, laptops, etc., on the other hand, the developed world will switch over to energy efficient infrastructure and consumer goods. Conversion of energy intensive to energy efficient system cannot be achieved overnight. It will take several decades if not centuries and till such time this happens, we will continue to enlarge human footprints.

The ongoing overconsumption and excess waste generation have forced humans to steal the future for present overlooking the cost to be paid by coming generations. The

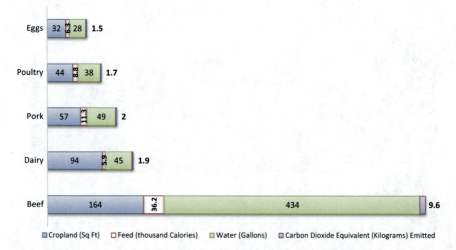

Fig. 4.6 American footprint consumption and emission to produce 1000 calories of energy for human consumption in the USA (*Source* Kunzig 2014)

humanity's ecological footprint[3] has been on the rise as a result of disproportionate increase in population, high densities and enormous increase in per capita energy. The ecological footprints of humans no longer coincide with their geographic locations. Most of the urban populations, for example, appropriate their water and electrical energy needs from distant sources and use industrial products through commercial trade from all over the world. During the period between 1970 and 1996, the global ecological footprint rose from 11,000 million hectares to 16,000 million hectares (Clarke et al. 2002). Recent estimates suggest economically rich countries and regions have larger footprints (Fig. 4.6) as compared to economically weaker nations and regions. For instance, the North America's ecological footprint was just over nine global hectares per person; that of western Europe is about five global hectares per person; while that of Asia-Pacific and Africa is around 1.5 global hectares per person during the latter half of last decade (Hails et al. 2008). The figures nine, five and 1.5 clearly reflect the huge consumption gap between the more developed and the less developed bipolar world.

Urban areas consume more energy and have therefore larger footprints than rural areas. The richest 20% in the world consume 86% of all goods and services and discharge 53% of global CO_2 emission as compared to 3% CO_2 generated by the

---

[3]Ecological Footprint—Area of terrestrial and aquatic ecosystems required to produce the energy needs of a given population and assimilate the end products. The concept of 'Ecological Footprint' has emerged as an important tool to ascertain the extent of productive land and water used by an individual, a village, a city or a nation in order to produce what is consumed and to absorb the waste generated. It is a function of population size, average per capita consumption of resources and the technology used. Ecological footprint of humans is expressed as 'global hectares' divided into four consumption categories; carbon (home energy and transportation), food, housing, and goods and services.

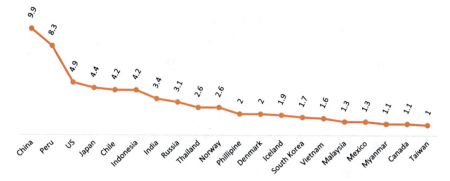

Fig. 4.7 Annual average landing between 2001 and 2005 (in million metric tons of fish) (*Source* Greenberg 2010)

poorest fifth who survive on 1.3% of all goods and services. The ecological impact of an average American is 30–50 times that of an average developing country citizen. The ecological footprint of London city, with 12% UK population and covering an area of 1700 km^2, was 21 million hectares or 125 times the surface area of London in 2001. Vancouver in Canada consumes the output of land area 174 times larger than its current size (Clarke et al. 2002). A typical North American city, with a population of 650,000 people, would require 30,000 km^2 of land to meet its domestic need (Munier 2005). At the current levels of consumption and waste generation, we need two to four Earths to normalize our footprints.

Demand for nutritious and healthy food has changed dramatically from 1950 onwards and developed world has switched over to large-scale consumption of marine animals. China's massive population gives it the world's largest seafood print with 694 million metric tons of[4] primary production followed by Japan at 582 million metric tons and America at 348.5 million metric tons (Greenberg 2010). During the past 50 years, the annual seafood catch has gone up by four times thanks to human demand, preference (or compulsion), new technologies and better fishing fleets that help hunting in previously unexploited regions. The increased demand of apex marine predators has outstripped the primary production capacities of the exclusive economic zones, and therefore, previously unexploited and virgin fishing grounds have now come under global fishing. The annual average production of fish between 2001 and 2005 was nearly 60 million tons out of which 67% went for food and 33% for industrial purposes such as paints, cosmetics and animal feed (Fig. 4.7); (Fig. 4.8).

---

[4]Primary production means indirect consumption of phytoplankton and algae by human for each unit of marine predator (bluefin tuna or shark or whale) consumed.

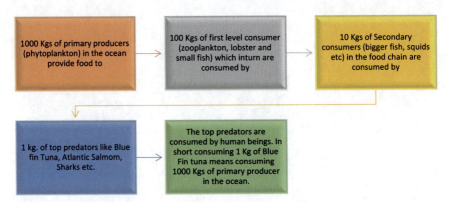

Fig. 4.8 Producer–consumer energy flow

> **Box 4.2 Bush Meat**
>
> Bush meat is a major food item that provides 20–90% of the total animal protein to most rural West Africans. Besides the rural population, the demand for and value of bush meat is rapidly increasing in many other regions because dwindling supplies from the wild.
>
> **Cameroon**–An estimated 2000 tons of bush meat are consumed annually in the country;
>
> **Ghana**–75% of the population regularly consumes wild animals mainly game meat, fish, insects, caterpillars, termites and snails;
>
> **Liberia**—Bush meat contributes 80–90% of the animal protein consumed. Small antelopes and monkeys are the popular species, and their population is declining;
>
> **Nigeria**—Wildlife is highly valued as food and 95% people consumed bush meat;
>
> **Senegal**—Wild life has become scarce due to consumption as staple food. (Source: Falconer 1990)

Increased consumption of ranched meat especially beef has become a source of fierce debate among the protagonists of climate change. Critics of industrial scale beef production are of the view that rearing cattle for beef not only warms climate, diverts land use, pollutes water but also deprives us of renewable animal energy. In 1976, the per capita beef consumption in USA was 91.5 lb per year that has come down to 54 lb a year now (Kunzig 2014). But US slaughters more than 8 billion chickens a year as compared to 33 million cattle, possibly to avoid mad cow scare and *E. coli* contamination (Kunzig 2014). In 2013, America produced 13 million tons of beef after feeding the animals with grains for 120–150 days. FAO has come to the conclusion that 6% of global GHG emission comes (Fig. 4.9) from beef and if the whole world abstained from beef, it will greatly help cutting down global warming

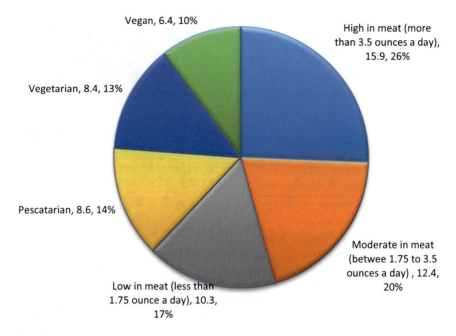

Fig. 4.9 Mean GHG emissions per 2000 kilo calorie diet of human being (in pounds of carbon dioxide equivalent) (*Source* Scarborough et al. 2014)

by reduced consumption of fossil fuel and fertilizers required for meat transport and feed crop.

Disproportionate production, consumption and wastage of farm produce by several countries for various reasons, including extreme weather conditions and poverty force millions to go hungry or remain malnourished. For example, erratic weather, recurring droughts and floods force two-thirds of Bolivians living in rural areas to rely on subsistence crops as a result of which 25% of children under 5 have stunted growth. Similarly, frequent natural disasters in Haiti including earthquake (in 2010), hurricanes (in 2012) and drought (in 2014) have aggravated food scarcity and poverty as a result of which 52% Haitians are undernourished. In such countries, people exert pressure on natural biological resources. Wild animals and plants supplement farm production, fill in seasonal shortfalls in food and provide buffer during the stress period. There are nearly 13 million naturally occurring biological species on this planet, and each one of them is a storehouse of irreplaceable substances.

Not all countries are biologically and economically rich enough to meet or supplement their nutritional requirements through ranched animals. For them, wild animals and plants supplement the farm production of millions of humans who live within or on the margins of forests, fill in seasonal shortfalls of food and provide buffer during the stress period. People rely on a range of animals including reptiles, birds and their eggs, bush buck, bush pig, grass cutter, deer and antelopes to meet their protein requirement. Both, marine and fresh water fish appear to be the most widely

Fig. 4.10 Number of plant species in medicinal use (*Source* Hamilton 2003)

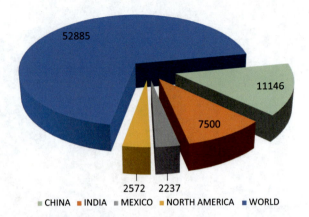

consumed followed by molluscs, shrimps, crabs and other invertebrates by the coastal communities around the world. Consumption of wild animals, however, varies from region to region depending on the degree of poverty. In West Africa, bush meat is still a major source of animal protein for poor villagers (Box 4.2). In Latin America, fish, invertebrates and mammals constitute a substantial portion of human diet. People living along the coast in Southeast Asia rely on fish, prawns and shrimps to a great extent. Meat of wild animals especially deer, antelopes, turtles, snakes, birds and fish is commonly consumed in many parts of South Asian countries (Fig. 4.10).

There are other reasons for overexploitation from wild. The techniques for artificial propagation of most wild plants have not been standardized as yet. Moreover, collection from wild is easy and cost effective for the traders who are aware that the expense involved in cultivation and time periods required for plants to reach marketable size make them unprofitable. Only 10% of the 1200–1300 species used in Europe are derived from cultivation. Even in China, despite its vast history of traditional medicine, only 100–150 species are cultivated. The situation is almost same in other countries of Asia as well as in Africa where more than 90% collection of medicinal plant is done from wild. Unsustainable harvesting and improper post-harvesting practices lead to over exploitation and resource depletion.

The increased current international demand coupled with other more serious proximate threats like habitat loss and habitat degradation is causing endangerment to biological diversity in the world. Unconfirmed statistics indicate that between 35,000 and 70,000 plant species are used as medicines and food supplements of which 9000 are threatened (McNeely and Sue 2006). Europe alone shares around 25% of world trade in medicinal plants, and the demand is growing.

Box 4.3 Nutritional value of Wild Food

1. Caldwell and Enoch (1972) found that, on an average, wild leaf vegetables have high riboflavin content (0.4–1.4-mg/100 gms edible portions);

2. R. Becker (1983) discovered that seed of chanar tree (Geoffroea decorticans) had a chemical score similar to that of groundnut and millet;
3. The vitamin C content of an orange is 57 mg/100 gms compared to baobab fruit's 360 mg/100 gms and Ziziphus jujuba's 1000 mg/100 gms;
4. Wild leaves, in addition to being cheap and accessible, can be an excellent source of:

   - Vitamin A—*Moringa oleifera* (11,300 μg/100 gms);
   - Vitamin C—*Cassia obtusifolia* (120 mg/100 gms.);
   - Calcium—*Balanites aegyptica* (37,010 mg/100 gms);
   - Niacin—Baobab (8.1 mg/100 gms);
   - Iron—*Leptadenia hastata* (95 mg/100 gms).

5. In Swaziland, 50% of the people consumed 48 species of wild leaves that they collected from agricultural fields, grazing and forestlands. 46% respondents reported buying them from local markets while 25% were engaged in selling;
6. Gum of Sterculia sps. is a dietary supplement and is a good source of vitamin C (52 mg/100 gms) and vitamin A (396 ug/100 mg retinol equivalent);
7. Gum arabic (Acacia senegal) is traditionally an important source of dietary supplement for hunter, gatherer, pastorals and nomads. Only six ounces is sufficient for a man per day;
8. 100 gms of honey provides 280 calories of energy.
(Source: Falconer et al. 1991)

4.3 Human Species Is Under Serious Threat

As population grows numerically and as the global economy expands, the shrinking per capita supply and availability of other natural biological and non-biological resources threaten the very existence of human species and the scale of this challenge is frightening as illustrated below:

- More than 83% Earth's land area is impacted by human activity endangering the existence of 25% wildlife and its habitat (Mayell 2002);
- In 19 of the 25 biodiversity-rich 'hot spots', population growth rates are higher (1.8% per year) than the global rate of 1.3% per year (Williams 2011);
- 16 of the 25 biodiversity-rich 'hotspots' are in areas where over 20% of human population is malnourished (Williams 2011);
- Of more than 17,000 major protected areas, 45% are heavily used for agriculture (Barber et al. 2004);
- Close to one billion rural poor live under chronic poverty and are forced to deplete natural assets;

- On average, 16 million people face starvation each year due to civil conflicts and natural disasters (FAO, IFAD and WFP 2014);
- Over 500 million people live in dry lands and over 625 million live in mountains inhospitable conditions;

The present skewed distribution of population will have following implications for the natural resources:

i. Pressures on natural resources will intensify in developing world due to high proportion of poor population.
ii. Migration of people will shift the pressure on natural resources depending upon population concentration. More megacities will come up along the coast with increasing population density as younger age group will migrate from mountains to plains and coasts.
iii. Developing economies will exert more pressure on natural resources through intensified resource consumption and production of waste.
iv. Urbanization and infrastructure development will increase the pollution levels in poor developing countries.

Environmental degradation and growing pollution are already threatening biodiversity and ecosystem stability, and many Asian countries are currently at a stage when their physical and biological systems may not be able to meet even their basic needs (potable water, fodder, energy, shelter and food) of the growing population. Increasing frequency of landslides, floods, droughts, soil loss and other human-incited natural disasters compound the miseries of poor communities' particularly marginal and poor farmers as well as the landless labourers. The true extent of habitat changes and species loss in the region is difficult to quantify in ecological as well as financial terms due to inadequacy of reliable data. The available data suggests that about two-thirds of Asian wild habitats have been destroyed and 70% of the major vegetation types in the South Asia, the Mekong basin and Southeast Asia have been lost, with a possible associated loss of up to 15% of terrestrial species (UNEP 2013). Both dry and moist forests have suffered losses as high as 70%. Not only that, wetlands, marsh and mangroves have been reduced by 55% (Braatz 1992).

Habitat losses have been most severe in China, Vietnam, Indonesia, Malaysia and Thailand. The 'hot spots' where the disappearance of already-threatened moist tropical forest would cause the greatest losses of biodiversity include the remaining forests in the Philippines, peninsular Malaysia, north-western Borneo, the eastern Himalayas, the Western Ghats in India, and the south-eastern Sri Lanka (Dowling 2008). A 'business-as-usual' scenario suggests that continued rapid economic growth and industrialization may result in further environmental damage and that the region may become more degraded, less forested, more polluted and less ecologically diverse in the future.

Humanity today dominates more than 50% of terrestrial biological production through agriculture, forestry and other activities and in doing so they have decreased the ability of global ecosystems to absorb the waste and store carbon. Disproportionately increasing human ecological footprint has usurped the space required for the survival and growth of thirteen million biological species which are currently confined to a mere 12% area, according to the World Commission on Environment and Development. Excluding this 12% area for biodiversity preservation, it can be inferred that out of approximately 2 ha per capita of biologically productive area on Earth, a mere 1.5 ha per capita[5] is available for human use (Sandhu et al. 1997). This 1.5 ha is the ecological benchmark for comparing people's ecological footprints. In case some individuals use more area (which is realistically possible), others will be compelled to use less area than entitled. Even if we believe that there will be no further ecological degradation and destruction (which is highly unlikely), the amount of available biologically productive space will drop to 1 ha per capita once the world population reaches its predicted number of 9 billion in 2050 (Wackernagel and Yount 1998). Post Industrial Revolution in general and twentieth century in particular has witnessed an unprecedented rise in human population, economic activity and social disintegration. The Earth's tolerance of expanded human activity has been overstretched, and the cumulative impact of world human population on natural ecosystems exceeds the sustainable yield capacity by nearly 30%.[6] Studies show that humanity has already depleted two-thirds of ocean fisheries and have irreversibly transformed half of Earth's terrestrial surface. That aggregate global output of goods and services (Gross World Output) increased from 6.7 trillion USD in 1950 to 48 trillion USD in 2002 is a reflection of overuse of natural biological and non-biological resources. Nearly one-fourth of our terrestrial ecosystems has been converted to farming through human actions. About 75% of the ecosystem services provided by nature have been on decline globally, and we are currently living on borrowed time (Mohan 2015). For example, by using up supplies of groundwater much faster than they can be recharged, we have depleted water sources and deprived our future generations. Unfortunately, the cost of such actions is borne by people far away from those enjoying the benefits of natural services (Kowalski 2013). The millennium report cautions that the increasing demand of growing human population will further exert greater pressures on the natural infrastructure, and it is time humanity realizes the seriousness of natural debt burden and prevents it from reaching the meltdown point.

4.3.1 Water Distresses

Water is the lifeline of our planet, and so long as the water sources are not managed effectively we shall continue to face serious environmental problems such as increased desertification, water logging, salinity, recurring droughts and floods, shortage of water for irrigation, industrial and domestic use and pollution.

---

[5] 1.5 ha of land per global citizen has been arrived at as follows (=0.25 ha of arable land + 0.6 ha of pasture + 0.6 ha of forest + 0.03 ha of built-up land).

[6] Natural ecosystems need to regenerate 30% more in a given time frame to ensure ecological sustainability.

Nearly 98% of all the liquid consumable water sources that exist at any point of time are in the form of soil moisture and groundwater. The remaining 2% are in the form of surface water in rivers, lakes and swamps. No justification is therefore required to ensure that the water percolation and water retention capacity of the soil is ever compromised. The interactions of physical, biological and chemical components of soil, water and vegetation perform extremely important ecological functions of:

- Water storage and purification;
- Retention of nutrients, sediments, pollutants;
- Storm protection and flood mitigation;
- Shoreline stabilization and erosion control;
- Groundwater recharge and discharge;
- Stabilization of local climatic conditions, particularly rainfall and temperature.

The demand for potable water will become one of the most pressing needs of the twenty-first century as the water consumption has risen several times compared to population growth. In many urban areas especially those with high population density, surface as well as groundwater withdrawals are so high that surface water supplies are literally shrinking, and groundwater reserves are being depleted faster than they have been replenished. Water sources in rapidly advancing (industrially) developing countries are now faced with serious problems of heavy metal and organic pollutants, eutrophication, acidification, while they continue to struggle with traditional problems of poor water supply and lack of sanitation services. Groundwater resources in most Asian countries are experiencing rapid growth in the mining and manufacturing sectors, the two big sources of groundwater contamination. It is estimated that global water situation will get considerably worse over the next 50 years if not earlier. Biologically rich water sources such as rivers and lakes that provide around 6% of the world's fish catch, or 7 million metric tons per year (FAO Fisheries and Aquaculture Department 2010) will be destroyed in near future if restorative actions are not taken.

Water facts:

1. While potable water supplies are abundant but its uneven distribution among and within countries causes serious water shortages, short-changing human water needs and damaging aquatic ecosystems;
2. As population rises, the water use in agriculture is also bound to increase to meet the rising demand for food. Agriculture already accounts for about 70% of water consumption worldwide, and the UN has estimated a 50–100% increase in irrigation water by 2025. Much of this projected increase will occur in developing countries, where population growth coupled with industrial and agriculture expansion will be the greatest;
3. Countries like China, India, North Africa, Saudi Arabia and the USA together pump more than 160 billion tons of aquifer water per year to produce only 160 million tons of grain;
4. China's water table is falling by 1.6 m per year and may find it difficult to feed its population in the coming decades (Brown 2001);

5. Water pollution further aggravates the existing scarcity of local and regional water supply by removing large volumes of water from the available supply;
6. Water quality in most of the developed countries has steadily improved due to strict legislation and major investments in new water and sanitation infrastructure. However, in poor nations where even availability of drinking water is scarce, enactment of such legislation remains mere paper tiger and contamination of usable water supplies remains a serious threat;
7. As clean water supplies diminish, competition between expanding urban and rural users increases. In those countries, where systems of water law and allocation exist, water markets can operate to transfer supplies between buyers and sellers for an agreed price. However, water pricing remains a highly sensitive issue in low-income countries, where most people depend on irrigated agriculture for their living;
8. In 1950, there were 5270 large dams but today there are more than 36,500. The number of waterways altered for navigation has grown from almost 9000 in 1990 to more than 500,000 impacting their viability as aquatic habitat;
9. The benefits of waterways to shipping, agriculture, dams, power generation and channelization projects remain important components of national development strategies, even though their adverse environmental impacts are well known;
10. Overfishing: Fishing happens to be the main source of livelihood and food in many developing worlds, and increased poverty makes it the most essential means of survival thereby driving many species towards extinction.
11. Water diversion: Increased demand for water by growing population and agriculture has made many developing countries to divert water for agriculture, industrial and urban supplies.

4.3.2 Risk to Marine Ecosystem and Corals

Among the terrestrial ecosystems, the most diverse are tropical rain forests and this is attributed to the abundance of insect species. Coral reefs, on the other hand, are the marine equivalent of the tropical rain forests in terms of productivity and biodiversity. Many experts consider coral reefs to be more diverse as it is estimated that the reefs have about 1 million species living in its extensive network of crevices or along the reefs. There are 1–2 million algae cells per square cm of coral tissue that live in symbiotic relationship with the polyps. Coral reefs, usually found in shallow seas of the tropics between the tropics of Cancer and Capricorn, occupy mere 0.17% of the total ocean flow but provide 12% of global marine fish catch. Besides, they also provide protection to cities and infrastructure along the coast, tourism and recreation benefits to local communities. The reefs built by coral polyps act as a carbon sink that removes carbon dioxide from the atmosphere. These polyps are very small, and it is estimated that 80,000 polyps make 1 kg of coral reef material. Coral reefs are currently confined to Australia (17%), Indonesia (16%), Philippines (9%), Hawaii (2%) and rest of the world (58%) (Burke et al. 2011).

Coral polyps are small flower-shaped animals which grow in millions forming a thin layer of living tissue on the surface of the reef. These polyps use calcium ions and carbonate ions for manufacturing skeleton of calcium carbonate. The polyps house symbiotic algae in their cells to capture sunlight and synthesize food for polyps which in turn secrete calcium carbonate or limestone. Corals colonies grow best in shallow and clear water with plenty of sunshine to support photosynthesis. Many species of coral also engage in 'mass spawning' whereby every polyp releases, in water, a pink sac that contains a sperm and an egg. A lower pH leads to decline in fertilization, in larval development and in attachment of larvae to substratum for producing new colonies. Acidification of the ocean also makes these corals vulnerable to corrosion especially the fast-growing branching corals and calcium secreting algae that help bind the reef. Failure to form new reefs also means loss of incredible biodiversity of millions of marine species including clams, cucumbers, squirts, fish, turtle and anemones.

Global warming due to increase in human population and increased economic activity over the past 50 years or more has resulted in increased pressures on the coral reefs. It is estimated that more than half a billion people living within 100 km of coral reefs are a serious threat to nearly 50% of the world's reefs (Veron et al. 2009). According to a study conducted by National Oceanographic Atmospheric Administration, USA, a warm pool of water started forming in the southern Indian Ocean at the end of December 1997 migrated west and north and crossed equator in April 1998 and entered northern Indian Ocean in May (Cinner et al. 2012). The result was mass bleaching and mortality of corals throughout the region. The rising water temperatures, coupled with coastal pollution, overfishing and coastal development damaged 16% of the world's coral reefs killing the symbiotic algae that provide 90% food to the polyps. Nearly 60 countries reported bleaching of corals in the year 1998 which included countries of West Asia, East Africa, Indian Ocean, South and Southeast Asia and Far East and Far East Pacific, the Caribbean and the Atlantic Ocean. The Indian Ocean was estimated to be most severely impacted, and more than 70% mortality was observed off the coasts of Kenya, the Maldives and the Andaman and Lakshadweep Islands. About 75% of the corals were reported to be dead in Seychelles Marine Park System and the Mafia Marine Park of Tanzania (Kamukuru et al. 2004).

Other major sources of damage to coral reef in the shallow coastal regions are the silt brought down by soil erosion due to the deforestation and construction; encroachment, agriculture, excess sedimentation due to poor land use practices, industrial effluents, urban sewage and fishing, clear felling in coastal forests and mining of coral rocks for building material, extraction of coral sands for cement production and collection of rare species for display, souvenir's and for aphrodisiac. These have brought large quantities of nutrients into the coastal waters resulting in phyto plankton bloom causing serious threats to coral reefs.

Currently, 10% of the coral reefs have already been reduced to skeletons, 30% are in critical condition and another 30% are under severe environmental stress. Scientists fear that if the trend in global warming continues, most of the corals will be dead by 2050 (Hoegh-Guldberg et al. 2007). The Great Barrier Reef system stretching

1400 miles is the most threatened is the most threatened due to climate change, temperature rise, acidification and cyclones. A study led by the Australian Institute of Marine Sciences found that between 1985 and 2012, the Great Barrier Reef lost 50% of its polyps risking the survival of 1500 fish species and 400 coral species.

One of the common pollutants of the marine ecosystem is sewage—raw or treated—that causes eutrophication when in over abundant quantity. The excess nutrients mainly nitrogen and phosphates lead to prolific breeding of minute plants near the sea surface, which prevents the sunlight from reaching down, reducing or even stopping photosynthesis, thereby leading to oxygen depletion in the sea. As a result, fish and other marine invertebrates die in large number. The danger of sewage lies in harmful bacteria that cause typhoid, dysentery, diarrhoea, cholera, etc. Although the bacteria causing such harmful diseases do not survive for long in sea water, they are taken in by animals like oysters and clams, which are subsequently, consumed by human beings. The sewage also contains man-made chemicals like pesticides or heavy metals, which get more or more concentrated as one moves up along the food chain. For example, fish growing in sea water with just 0.1 part of DDT per billion parts water will have 57 mg. DDT per kg for body weight. This increases to 800 mg/kg in sea gulls, which feed on these fish. The eggs laid by seagulls have thin shells, which break during incubation by the mother. Unfortunately, DDT has reached even remote places like the Arctic Circle and the Antarctica and has been found in large concentrations in the bodies of polar bears, seals and penguins. In addition, there are other pesticides like aldrin, dieeldrin and endrin that were abundantly used 2–3 decades ago but have not been easily biodegraded. Heavy metals like chromium, lead, cadmium, mercury and nickel are fatal to marine fauna. A classic example is the Minamata disease, named after a place in Japan where the factory was dumping mercury waste into the sea. Fortunately, only 43 out of 50,000 inhabitants died due to impact of heavy metals.

4.3.3 Changing Temperatures and Forests

Warmer air, changing weather pattern and declining rainfall have created drought like conditions in many parts of the world in recent times. Warm air reduces the availability of water for the trees in forests, farms and plantations on the one hand and increases loss of moisture through transpiration and soil moisture loss on the other. This unprecedented combination of drought and heat has resulted in massive loss of trees and vegetation in all parts of the world be it the North or South America, south-west Australia, Asia or Mediterranean region. It has altered the ecosystem dynamic. For example, in British Columbia and the north-west parts of the USA, the mountain pine beetle (Fig. 4.11) whose primary host has been lodge pole pine had been, for centuries, living in mutually beneficial relationship by attacking dead, dying and weaker trees of lodge pole and ponderosa pine and producing humus as well as more beetles in return. Cold temperatures kept their population under check, and ecosystem energetic was well maintained (Rosner 2015). Their life cycle

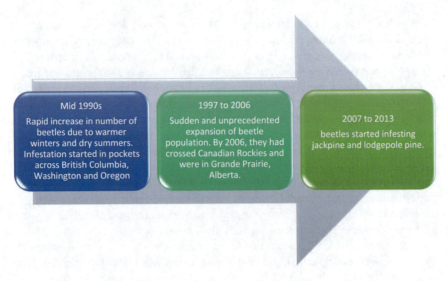

Fig. 4.11 Beetle infestation in North America

would begin in summer when the female will lay eggs under the bark of dead, dying or diseased trees. Healthy trees, if attacked, normally release toxins that annihilate beetle population.

However, warmer weather conditions and droughts have increased the beetle population and range. They have, since mid-1990s, spread northwards and eastwards across drought stricken British Columbia and the states of Washington, Idaho, Colorado, etc., destroying trees in more than 60 million acres over past 25 years leading to massive loss of timber (around one billion cubic metre), loss of habitat and food for birds and animals, loss of soil and massive forest fires. Not only this, the beetles have now invaded other species of pine such as jack pine and whitebark pine. Earlier, these beetles would invade in the month of July hopping or flying from one tree to another and increasing their population, but now their breeding season lasts up to October which means more trees to attack and expansion of its area of operation.

An effective strategy to contain the spread of beetle is to confine them to their present range. This can be done by cutting down large number of trees and burning them. But this means spending millions of dollars, increasing CO_2 and other gases and changing the whole energy dynamics in the region.

4.3.4 Rising Seas

> *If 5 trillion tons of carbon is added by burning all coal, oil and gas, we are very likely to create an ice free planet. Because then the average temperature will be 80 °F instead of 58 °F now and the sea level would be 216 ft. higher than the current level* (McKibben 2012).

The sea level or the height of ocean surface is ascertained in two ways. The Relative Sea Level (RSL) is the vertical motion of the sea or land surface that causes changes in sea level relative to the land. The other is Geocentric Sea Level (GSL) which is the height of the ocean surface relative to the centre of the Earth. The RSL is measured through tide gauges installed at thousands of ports across the globe, and these gauges provide temporal data ranging from recent to as old as eighteenth century. Measuring GSL is a recent phenomenon with the use of satellite altimetry. Mean Sea Level, a term commonly used in climate change science, is the average of sea level measured over a period of time to remove short-term variability in sea level at a particular place.

Under normal circumstances, exchange of water and ice between land and sea, rainfall on land surface, ocean current, ocean density, temperature and salinity contribute to and adjust sea level changes in a dynamic way. What has changed over the years is that the natural flow of water from land to sea has been obstructed by the dams and reservoirs as well as changes in surface run-off, percolation and evapotranspiration. This may result in sea level fluctuations at many places. Secondly, most of the excess heat, due to global warming, is absorbed in the upper 700 m of the ocean causing thermal expansion of the ocean. This thermal expansion is responsible for 40% of the observed sea level rise since 1970. Thirdly, nearly 50–60% sea rise comes from melting of ice on land. There are more than five million cubic miles or 8 million cubic kilometres of ice on Earth, and it may take several thousand years of warming to melt all of this. Currently, the largest reservoir of fresh water in the form of ice are the Greenland, West and East Antarctic ice sheets that gain and lose mass (water) through snowfall and melting. Greenland loses roughly half of its accumulated ice through surface melting and the other half by splitting of iceberg also known as calving. In East Antarctica, snowfall is expected to increase in twenty-first century mainly because warmer atmosphere will carry more moisture to polar region and the temperatures are too cold for surface melting. West Antarctica, on the other hand, may lose ice over the years as it most of the ice loss in Antarctica has so far come from West Antarctica Ice sheet.

Of nearly 734,000 km^2 of global glacier area exclusive of Greenland and Antarctic ice sheets, nearly 280,500 km^2 drains through marine terminating outlets (Hill et al. 2017). Glaciers are receding faster than expected—Columbia glacier in Alaska 7.65 gigaton per year between 1996 and 2007 (Church et al. 2013). Glacier National Park had 150 glaciers a century ago, but most of these either retreated or disappeared altogether (Diolaiuti and Claudio 2010). Studies by US Geological Survey revealed that by 2010 only 37 glaciers remained out of which only 25 were considered as 'active' meaning they were of 0.1 km^2 in area. If the current global warming trend persists, even these 25 will disappear by 2030.

Himalayas meaning the 'abode of snow' are the world's youngest, and highest mountain ranges that separate the Indian subcontinent from the Tibetan plateau and provide food and water security to billions of people living downstream. The Himalayan *glaciers* have been well acknowledged for the scenic beauty, and the movement of these large mass of ice under the force of gravity has created alpine wonderlands and glacial lakes. Seasonal melting of these glaciers produces the life force of fresh water for great rivers including the Ganges, the Brahmaputra and the Indus. These glaciers are a wonderful indicator of changing climate as their cycles of advance and retreat to changes in temperature serve as a valuable indicator of change within the climate system. Direct observation of the snouts of a few Himalayan glaciers indicates that they have been in a state of decline over the past 150 years. Melting glaciers are filling Himalayan mountain lakes and river systems too quickly, threatening millions of lives with flash floods and landslides. If the glacial retreat continues over the long term (several decades), as would be expected in a greenhouse-gas warmed climate, the amount of water will decrease and the flow of rivers in southern Asia become less reliable and eventually diminish, leading to potential widespread water shortages leading to disastrous consequences.

Arctic region that plays a crucial role in regulating Earth's climate is also shrinking at an unprecedented rate. Global warming in last several decades have accelerated the melting of glaciers in the Arctic and has also changed the number of days of sea ice cover. Instead of forming ice cover in December–January and melting in June, the sea freezes in February and starts melting in April. Not only that, climate change has affected the thickness of snow making it difficult to sledge or snow mobile to move in relatively thin ice cover. This in turn affects the hunting of whale, seal and reindeer for year-long requirement of meat and fur, etc. Floating arctic sea ice melts and refreezes annually, but the speed of this change is now twice as great as it was in 1972. Arctic temperatures have also risen more than twice as fast as the global average over the past half century. Scientists of the Polar Science Centre, University of Washington, are of the view that the arctic ice volume was lowest in 2010 about 2135 cubic miles—half the average and 62% lower than the maximum covering in 1979 (Stroeve et al. 2007). In the event of continued ice plunging, the Arctic will be free of ice in summer months in next 30–40 years, a state that was witnessed during last major interglacial period 125,000 years ago.

Satellite measurements of arctic sea ice since 1979 project a scary future for the entire ecosystem. From 7.19 million square kilometres of ice in 1979, the arctic ice was reduced to 4.63 million square kilometres in 2015 (Lauren et al. 2016), a massive shift due to climate change. Reduction in freeze time and increase in melt period exposes more water to sunlight and since open water absorbs 93% sunlight, it further reduces the freeze period in the ensuing freeze season (October–March). Reduction in freeze time has another implication-reduced albedo. Under normal circumstances arctic ice reflects 85% sunlight but as the snow reduces and ice sheet thin out, the albedo effect is also reduced. In other words, the current warming of Arctic ecosystem will have amplifying effect that may be difficult to retrieve.

Increased melt period and sunlight absorption have enhanced the Arctic Ocean surface temperature by more than 1 °F between 1985 and 2015 (Lauren et al. 2016) and the air above by 5 °F in the past century. Arctic water absorbs more carbon dioxide now, a sign of more ocean acidification and loss of carbonate, an indication that shell-forming arctic animals such as snails and crabs may lose out. Climate models now predict that by 2040, it will be possible to sail across open water to the North Pole, a sign of optimism for tourism industry and economist!

Impacts of warming in the Arctic over the next century will exceed the impacts forecast for many other regions and will react with globally significant consequences. This grim forecast is already proving real with thawing of permafrost, greening of land, northward shifting of tree line and invasion of shrubs and grasses in the tundra. This loss of ice cover may also impact marine animals including polar bear, killer whales, pacific walrus, seabirds, seals in a manner difficult to predict now.

Normally, ice in the Arctic is 10–13 ft. thick and perennial. But this has changed for worst now with thickness reduced by nearly half, and the extent of thin sheet has also increased reducing albedo effect.

4.3.5 Droughts and Fire

Central and Southwest Asia

Most of the SW Asia including Iran, Iraq, Afghanistan, Pakistan, Tajikistan, Uzbekistan was hit by a severe drought between 1998 and 2003 with less than 55% (of the long term average) average rainfall making 2000 as the worst drought in past 50 years. Some parts of Iran did not receive rains for 30 consecutive months. Persistence of this kind of extreme weather was attributed to the formation of anomalous ridge in the upper-level atmospheric circulation during the cold season in central and southern Asia (Jones et al. 2007). This inhibited the development of storm system on the one hand and deflected the eastward-propagating storms to the north of the drought area.

The Thirsty Australia

Australia, the most arid continent on the Earth, is repeatedly subjected to dry spells the worst being the seven year drought in the first decade of twenty-first century. Over the years, the soil has become infertile and the blame is largely attributed to the actions of early settlers who felled 15 billion trees in Murray-Darling Basin (Draper 2009). The natural water cycle was disrupted, and soil was impoverished with the introduction of sheep, cattle and water hungry crops. The semi-arid land of the basin was transformed into breadbasket of Australia through a massive water management programme that dammed rivers, filled reservoirs for irrigation and other purposes. Commercial farming of cotton used up most of the water followed by dairy farming, rice, grapes and other fruits and vegetables.

Australia has been repeatedly affected by severe droughts during the twentieth century and thereafter as well. 1982, 1994 and 2002–2003 have been remarkably bad years. The mean annual maximum temperature during 2002 was 0.5 °C warmer than in 1994 and 1 °C warmer than in 1982. The drought of 2002 continued for three consecutive years with deficit rainfall in the eastern parts of Australia.

Drought and heat conditions revisited Australia in 2013 when average temperature was 1.5 °C above the norm for 1911–40, and this was attributed to human influence on climate. Australian Government spent $4.5 billion between 2001 and 2012 as drought relief assistance to farmers and others (Steffen 2015).

Drought, Fire and Frost in Russia

2010 will go down in the history of Russia as the worst year that aggravated heatwaves, caused severe drought, burnt down hundreds of villages, left thousands homeless and triggered wildfires. Wild forests cover two-thirds of Russia's territory, and forest fires during summers are not uncommon. However, during the year 2010 when the temperature went as high as 40 °C, engulfing 800,000 ha in Central Russia forcing around 200,000 firefighters to contain and douse it. Thick smoke from the burning woods and peat blogs blanketed Moscow and other cities for several days creeping into houses, offices, stores and underground metros. Satellite images showed smoke rising 12 km above the Earth forming dense clouds leading to cancellation of 60,000 flights. The primary cause for the unprecedented wildfire was the hottest and driest summer (of 2010) in at least one thousand years. In addition to deaths and infrastructure destruction, the government was forced to remove all radioactive and explosive material from Federal Nuclear Centre near the town of Sarov in Central Russia, and missiles and other air defence equipment near Moscow. While some experts termed this as natural disaster aggravated by climate change, there were others who blamed large-scale uncontrollable fire to wrong policies of privatization because of which the state or its agencies could not intervene in forest or fire management (Radyuhin 2010).

Post 2011, Russia saw the harshest frost in fifty years in 2013 with temperature dropping to minus 45 °C in north and minus 25 °C in central parts of the country during the month of March. The region was covered under 70 cms of snow not experienced in last 20 years and causing death of many wild animals in Russia's Far East (Radyuhin 2013). The biting frost did not spare much of Europe, the USA, China and India from abnormally icy weather. Life in Kiev was brought to standstill for several days, and more than 100,000 homes in Poland had to fight freezing cold without electricity. Meteorologists attribute this abnormal freeze to atmospheric arctic blocs, a phenomenon resulting from rapid loss of arctic ice in recent years that caused higher than usual air pressures over the polar regions pushing cold currents southwards.

Drought in Western North America

For five consecutive years, from 1999 to 2004, the western USA, southern Canada and north-west Mexico suffered severe shortfall in rainfall leading to hydrologic stress, reduced streamflow, drying up of lakes and reservoirs and low soil moisture. The drought conditions were attributed to changes in atmospheric circulation associated with warming of western tropical Pacific and Indian oceans. Colorado River suffered five consecutive years of below average flow since 1922. In western USA, there was 20% increase in drought areas that continued for five years, reached its peak in August 2002 and affected 87% of the western region creating an extreme situation not seen for 105 years history of the region (Jones et al. 2007).

Drought in Western America

Four years of consecutives drought since 2011 has led to water crisis in California, Nevada, Arizona, Oklahoma, New Mexico and Texas. The Golden State of California fell short of 2 billion cubic metres of water in 2015 as the state's 1440 reservoirs dried up impacting 39 million residents (The Economist 2015b). The western region of USA is a naturally dry region and relies heavily on underground water and dams to provide water for irrigation and domestic use. Most of the water comes through winter storms from the Pacific Ocean that deposits snow on mountain ranges (Nijhuis 2014). Hundreds of dams have been constructed over the years to tap the melting water. Most of the river systems in the region are saddled with dams, canals and aqueducts to make uninhabitable land habitable, convert unirrigated land to irrigated ones and encourage industrial water use. The western region consumes nearly 200 gallons per capita per day, 50 gallons more than the rest of USA. The agriculture and domestic water use close to 90 billion gallons a day in west which is almost double the quantity used by rest of USA. Farmers in the western region use 80% of state water to grow high water-consuming crops such as walnut, almond, tomatoes, broccoli and grapes in thousands of acres have been depleting underground water sources to irrigate their crop fruits, nuts and vegetables. As a result of persistent drought, the State of California imposed water cut by 25% for all its residents declining per capita use of 232 gallons in 1990 to less than 170 gallons now.

Wildfires have become more frequent and intense especially Ponderosa pine and Chaparral vegetation around south-west coast of California, south Idaho, north Nevada, west Mexico, etc., increasing local temperatures and further drying up of water sources.

Scientific predictions are that, in future, the south-western region will receive less precipitation vis-a-vis north-western. Declining trend in snowfall over the years is likely to continue, and warmer winters are reducing the amount of snow stored in the mountains. The region has been losing four trillion gallons of water every year from the Sacramento and San Joaquin river basins. 75% of this lost water comes from the aquifers in the Central Valley where two large sunken bowls have been formed threatening dams, bridges and waterways. Snow has virtually disappeared from Sierra Nevada, a crucial water reservoir.

Heatwave and Floods in Europe
The citizens of Europe will never forget the blistering summer of 2003 when the absolute maximum temperature exceeded the highest temperatures ever recorded since 1780, the year when temperature recording began. The mercury went up by 3.8 °C above the 1961–1990 average (The Economist 2015a). This heatwave was attributed to a very robust and persistent high-pressure system possibly a manifestation of northward extension of the Hadley Cell. This was exacerbated by absence of rainfall in many parts of western and central Europe. Strangely, many central European rivers suffered catastrophic floods in the preceding year due to extraordinarily high precipitation. The water levels in Elbe River in Dresden reached 9.4 m, the highest mark since 1275.

Yellowstone National Park, USA 1988
The first national park of the world suffered worst drought in the recorded history of 112 years. A total of 249 cases of fire were reported inside the park damaging more than 211,000 ha of natural vegetation.

Canada: Forest Fires of 1989

The Manitoba Province of Canada was devastated by massive fire in 1989 which erupted from the shores of Lake Manitoba and spread to more than thousand places northwards for 800 km. Huge flames engulfed a kilometre zone across the Nelson river, forcing the evacuation of 4000 people from the northern community of Norway House and cut off rail and road transport links. The reason for this fire was primarily high temperatures and drought over the previous two years that dried out many peat bogs and wetlands and facilitated spread of fires. The inaccessibility of the area and large number of the fires made it difficult to deal with the situation. By autumn, Manitoba had been hit by 1140 fires that spread to 2.7 million hectares covering 5% of Manitoba's land area. More than 25,000 people from dozens of northern towns were displaced and became refugees in their homeland. Similar huge fires raged east of James Bay in Quebec and smaller fires struck Saskatchewan and Ontario. By the end of the year, about 6.4 million hectares of forest was destroyed.

Vulnerable to Sea Rise
Florida

National Climate Assessment of the US Government predicts doomsdays for Florida in 2100 by heat, droughts and deluges. 75% of the Florida's 18 million population resides along the coast and are responsible for generating 80% of their economy. Sounds great but the downside of this economic development is that 50% of 825 miles of beaches are eroding fast as buildings, roads and bridges come up fast to improve two trillion dollar economy of the state. According to the scientific assessment, if the Greenland and polar ice sheets melt, the ocean could rise by two feet by 2060 and with every foot rise, the shoreline would move inland by 500–2000 ft.

Bangladesh

With 164 million people in 2011, Bangladesh is one of the most densely populated countries. From the vantage point in the Bay of Bengal, the country is already witnessing sea level rise, salinity ingress and frequent cyclones. Two of its major cities, Dhaka and Chittagong are most vulnerable to coastal flooding and rise in sea levels in the near future. Nearly 10–30 million people will be uprooted from the southern coast and will seek refuge in India, north Bangladesh, Burma and other countries by 2100.

Kiribati

The island nation of Kiribati, covering 310 square miles of land and 1.3 million square miles of ocean, is a collection of 33 coral islands in the central Pacific Ocean. Thirty-two of the islands are low-lying atolls and the 33rd called Banaba is a raised one that was strip mined for seabird guano-derived phosphates long ago. If scientists are to be believed, all 33 islands are highly vulnerable to rising sea, acidification, coastal erosion, salinity ingress, unusually high tides, flooding. Wells are becoming brackish, and agriculture crops and trees are fast disappearing. The IPCC has predicted that the sea levels will rise by nearly 1.5 foot by 2100 as the coral island is severely affected by Pacific decadal oscillation (due to changes between the El Nino and La Nina). In March 2015, Kiribati experienced flooding and massive destruction of coastal properties due to cyclonic storm (Cyclone Pam) that not only devastated Vanuatu Island but also affected Solomon Island, Tuvalu and New Zealand. The bleaching of coral reefs has become more frequent threatening the survival of coral itself. And if and when the corals go, the island will follow soon.

Pollution Deaths in Europe

More than 400,000 people die premature deaths each year due to air pollution in Europe. 90% of European city dwellers are exposed to high levels of nitrogen oxides, sulphur dioxide and particulate matter every year. In particular, the residents of London are exposed to high levels of nitrogen dioxide; several cities in Turkey are choked with high levels of PM10 (particulate matter of maximum 10 μm diameter) which can change the immune system of lungs especially in children and stunted foetal growth.

A recent report by Bulletin of the American Meteorological Society on nine different studies including in Europe, China, Japan and Korea showed that manmade climate change has increased the likelihood of exception heat. In Korea, daily minimum summer temperatures were 2.2 °C above 1971–2000 average; in Germany, the probability of exceptional heat is once in 7 years now as compared to once in 80 years during pre-industrial phase (The Economist 2015a).

The Federal Institute of Technology in Zurich analysed the heat and precipitation extremes between 1901 and 2005 (extremes events were those that recurred once every 1000 days) and concluded that 0.85 °C rise in temperature has made such heat extremes four or five times more likely. Also the probability of heat-related extremes is twice as great at 2 °C than at 1.5 °C (Schiermeier 2008).

References

Barber, Charles Victor, Melissa M. Boness., and Kenton Miller, (eds.). 2004. *Securing protected areas in the face of global change: Issues and strategies*. World Commission on Protected Areas, IUCN–the World Conservation Union.

Becker R. 1983. Nutritional quality of the fruit from the chañar tree (Geoffroea decorticans). *Ecology of Food and Nutrition*. 1 Apr 1983;13(2):91–7.

Binford, Michael W., Alan L. Kolata, Mark Brenner, John W. Janusek, Matthew T. Seddon, Mark Abbott, and Jason H. Curtis. 1997. Climate variation and the rise and fall of an Andean civilization. *Quaternary Research* 47 (2): 235–248.

Braatz SM. 1992. Conserving biological diversity: a strategy for protected areas in the Asia-Pacific region. World Bank Publications

Bradshaw, Corey JA., and Barry W. Brook. 2014. Human population reduction is not a quick fix for environmental problems. *Proceedings of the National Academy of Sciences* 111 (46): 16610–16615.

Brown, Lester Russell. 2001. *State of the world, 2001: A worldwatch institute report on progress toward a sustainable society*. WW Norton & Company.

Burke, Lauretta, Kathleen Reytar, Mark Spalding., and Allison Perry. *Reefs at risk revisited*. 2011.

Caldwell MJ, Enoch IC. 1972. Riboflavin content of Malaysian leaf vegetables. *Ecology of food and nutrition*. 1 Sep 1972; 1(4):309–12.

Church, John A., Peter U. Clark, Anny Cazenave, Jonathan M. Gregory, Svetlana Jevrejeva, Anders Levermann, Mark A. Merrifield et al. 2013. Sea-level rise by 2100. *Science* 342 (6165): 1445–1445.

Cinner, Joshua E., T. R. McClanahan, N. A. J. Graham, T. M. Daw, Joseph Maina, S. M. Stead, Andrew Wamukota, Katrina Brown., and Ö. Bodin. 2012. Vulnerability of coastal communities to key impacts of climate change on coral reef fisheries. *Global Environmental Change* 22 (1): 12–20.

Clarke, Robin, Robert Lamb., and Dilys Roe Ward. 2002. *Global environment outlook 3: past, present and future perspectives.* Earthscan.

Cohen, Joel E. 1995. How many people can the earth support? *The Sciences* 35 (6): 18–23.

Diolaiuti, Guglielmina., and Claudio Smiraglia. 2010. Changing glaciers in a changing climate: How vanishing geomorphosites have been driving deep changes in mountain landscapes and environments. *Géomorphologie: relief, processus, environnement* 16 (2): 131–152.

Dowling, John Malcolm. 2008. *Future perspectives on the economic development of Asia,* vol. 5. World Scientific.

Draper, Robert. 2009. The climate betrayed him-Australia's dry run. *National Geographic* 215: 36–75.

Ehrlich, Paul. 1970. The population bomb. *New York Times,* 47.

Falconer, Julia. 1990. *The major significance of minor forest products.* Rome, Italy: FAO.

Falconer, Julia., and JE Michael Arnold. 1991. *Household food security and forestry: An analysis of socio-economic issues.* Rome, Italy: Food and Agriculture Organization of the United Nations.

FAO. The State of Food Insecurity in the World 2008. High Food Prices and Food Security–Threats and Opportunities.

FAO Fisheries and Aquaculture Department, "The state of world fisheries and aquaculture." *Food and Agriculture Organization of the United Nations, Rome* (2010).

FAO, IFAD., and WFP, The State of Food Insecurity in the World 2014. Strengthening the enabling environment for food security and nutrition, 2014. Rome: FAO.

Greenberg, Paul. 2010. Time for a sea change. *National Geographic* 218: 70–89.

Hails, Chris, Sarah Humphrey, Jonathan Loh, Steven Goldfinger, Ashok Chapagain, G. Bourne, R. Mott et al. 2008. Living planet report 2008.

Hallegatte, Stephane, Colin Green, Robert J. Nicholls, and Jan Corfee-Morlot. 2013. Future flood losses in major coastal cities. *Nature Climate Change* 3 (9): 802.

Hamilton, Alan. 2003. Medicinal plants and conservation: Issues and approaches. *International Plants Conservation Unit, WWF-UK* 51.

Haub, Carl., and Diana Cornelius. 2007. *World population data sheet* Washington, DC: Population Reference Bureau.

Hawksworth, John., and Gordon Cookson. 2006. The world in 2050. *How big will the major emerging market economies get and how can the OECD compete.*

Hill, Emily A., J. Rachel Carr., and Chris R. Stokes. 2017. A review of recent changes in major marine-terminating outlet glaciers in northern Greenland. *Frontiers in Earth Science* 4: 111.

Hoegh-Guldberg, Ove, Peter J. Mumby, Anthony J. Hooten, Robert S. Steneck, Paul Greenfield, Edgardo Gomez, C. Drew Harvell et al. 2007. Coral reefs under rapid climate change and ocean acidification. *Science* 318 (5857): 1737–1742.

Hunter, Lori M. 2000. *The environmental implications of population dynamics.* Rand Corporation.

Jones, P. D., K. E. Trenberth, P. Ambenje, R. Bojariu, D. Easterling, T. Klein, D. Parker, J. Renwick, M. Rusticucci., and B. Soden. 2007. Observations: Surface and atmospheric climate change. *Climate Change*: 235–336.

Kamukuru, Albogast T., Yunus D. Mgaya, and Marcus C. Öhman. 2004. Evaluating a marine protected area in a developing country: Mafia Island Marine Park, Tanzania. *Ocean and Coastal Management* 47 (7–8): 321–337.

Kowalski, R. 2013. Sense and sustainability: The paradoxes that sustain. *World Futures* 69 (2): 75–88.

Kunzig, Robert. 2011. Population 7 billion. *National Geographic* 219 (1): 32–63.

References

Kunzig, Robert. 2014. Carnivore's Dilemma. *National Geographic*: 108–131.
Kolbert, Elizabeth. 2009. Outlook: Extreme. *National Geographic*.
Lauren, E. James, Jasen, Treat., and Williams, Ryan et al. 2016. Disappearing ice. *National Geographic* 3 (6).
Mayell, Hillary. 2002. Human "footprint" seen on 83 Percent of Earth's Land'. *National Geographic News October* 25 (2002).
McKibben, Bill. 2012. Global warming's terrifying new math. *Rolling Stone* 19 (7): 2012.
McNeely, Jeffrey A., and Sue Mainka. 2006. The future of medicinal biodiversity. *IUCN*, 165.
Miller, Peter. 2012. Weather gone wild-disastrous rains. No rain at all. Unexpected heat or cold. Is Earth's climate changing dangerously? *National Geographic* 222 (3): 30.
Mohan, Brij, ed. 2015. *Global frontiers of social development in theory and practice: Climate, economy, and justice*. Springer.
Munier, Nolberto. 2005. *Introduction to sustainability: Road to a better future*. Springer Science & Business Media.
Nijhuis, Michelle. 2014. When the snows fail. *National Geographic* 226: 58–77.
Radyuhin, Vladimir. 2010. Russia: Forest up in flames. *The Hindu*, Newspaper.
Radyuhin, Vladimir. 2013. Down to minus 45. *The Hindu*, Newspaper.
Rosner, Hillary. 2015. The bug that's eating the woods. *Natl. Geogr. Mag.*
Sandhu, Shabeg, Laura Jackson, Kay Austin, Jeffrey Hyland, Brian Melzian., and Kevin Summers. 1997. Monitoring ecological condition at regional scales.
Scarborough, Peter, Paul N. Appleby, Anja Mizdrak, Adam DM Briggs, Ruth C. Travis, Kathryn E. Bradbury., and Timothy J. Key. 2014. Dietary greenhouse gas emissions of meat-eaters, fish-eaters, vegetarians and vegans in the UK. *Climatic change* 125 (2): 179–192.
Schiermeier, Quirin. 2008. Global warming brews weird weather. Nature news. Aug 27th (2015).
Food and Agriculture Organization of the United Nations. *The State of Food Insecurity in the World 2008: High Food Prices and Food Security-Threats and Opportunities*. FAO.
Simon, Julian Lincoln. 1996. *The ultimate resource 2*. Princeton: Princeton University Press (revised in 1998).
Smith, Adam B., and Richard W. Katz. 2013. US billion-dollar weather and climate disasters: Data sources, trends, accuracy and biases. *Natural Hazards* 67 (2): 387–410.
Steffen, Will. 2015. *Thirsty country: Climate change and drought in Australia*. Sydney, Australia: Climate Change Council of Australia Ltd.
Stone, Brian, Jeremy J. Hess, and Howard Frumkin. 2010. Urban form and extreme heat events: Are sprawling cities more vulnerable to climate change than compact cities? *Environmental Health Perspectives* 118 (10): 1425.
Stroeve, Julienne, Marika M. Holland, Walt Meier, Ted Scambos., and Mark Serreze. 2007. Arctic sea ice decline: Faster than forecast. *Geophysical Research Letters* 34 (9).
The Economist. 2015a. Is it global warming or just the weather? *The Economist*, May 9th.
The Economist. 2015b. California's drought, All the leaves are brown. *The Economist*, May 28th.
The Economist. 2015c. Insurance in Asia-Astounding claims. *The Economist*, Jun 11th 2015c.
United Nations. 2001. Population, environment and development. The Concise Report.
United Nations Environment Programme (Unep). 2013. *Global environment outlook 2000*, vol. 1. Routledge.
Veron, J.E.N., O. Hoegh-Guldberg, T.M. Lenton, J.M. Lough, D.O. Obura, P. Pearce-Kelly, C.R.C. Sheppard, M. Spalding, M.G. Stafford-Smith, and A.D. Rogers. 2009. The coral reef crisis: The critical importance of <350 ppm CO_2. *Marine Pollution Bulletin* 58 (10): 1428–1436.
Wackernagel, Mathis., and J. David Yount. 1998. The ecological footprint: An indicator of progress toward regional sustainability. *Environmental Monitoring and Assessment* 51 (1–2): 511–529.
Williams, John N. 2011. Human population and the hotspots revisited: a 2010 assessment. In *Biodiversity hotspots*, 61–81. Springer, Berlin, Heidelberg.

Chapter 5
Reducing Carbon Growth

> *The total power need of the humans on Earth is approximately 16 terra watts which is expected to grow to 20 terra watts in 2020.*
>
> Johnson (2009).

Abstract Under the present circumstances of global warming, two actions are urgently required. One is to reduce the emission from greenhouse gases, and the other is to encourage climate engineering efforts such as carbon capture and storage and solar radiation management. GHG reduction must be done at all cost by all nations, whereas climate engineering technology can be introduced by rich countries at the earliest. Energy efficiency is one of the major energy-related threats being faced in most of the countries. The International Energy Agency estimates that in order to meet the energy efficiency target for 2050, the global spending has to increase from the current level of 300 billion USD per year to 680 billion USD. A business-as-usual scenario suggests that USA and European Union would reduce oil and coal consumption to improve energy efficiency by 2050. However, China and India, the two most populous nations will continue to rely heavily on coal and oil in near future and will not be able to transit to renewable source to the expected level. Like it or not, we will have to explore all possibilities of switching over to a combination of source from renewable energy basket including opting for draft animal power.

Keywords Solar radiation management (SRM)
Carbon capture and storage (CCS) · Energy efficiency · Solar energy
Wind energy · Geothermal energy · Lithium · Carbon intensity · Energy intensity

5.1 Introduction

There are many ways of cooling the ongoing Earth's warming besides reducing the emission of greenhouse gases which is, without doubt, the primary action that must

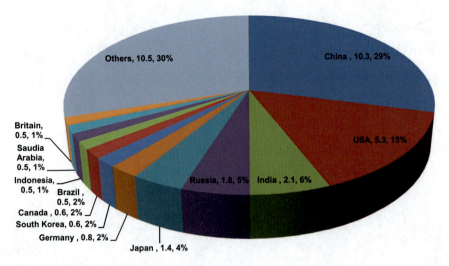

Fig. 5.1 Carbon dioxide emission in billion tons (2013). *Source* The Economist (2015a)

be taken soonest. Besides, the climate engineering efforts are seen as secondary measures that need to be dovetailed with emission reduction efforts. These measures can be divided into two categories, the management of carbon and the management of sunlight. The prominent example of carbon management includes carbon capture and storage (CCS) where some of the carbon dioxide emitted by coal-fired thermal plants is recaptured by physically sucking it in and transporting it elsewhere to be sequestered underground. For example, the 115 MW CCS retrofitted coal power plant at the Boundary Dam in Canada commenced operation in 2014 (Parfomak et al. 2007). The CO_2 captured in the plant is pumped into nearby oil fields for enhanced oil recovery.

The purpose of sunlight management or solar radiation management (SRM) is aimed at cutting down the heat absorbed by our planet from the Sun. One of the well-known techniques of SRM is called stratospheric aerosol injection (SAI) which involves spraying of fine, light-coloured particles into the stratosphere in order to reflect back part of the solar radiation before it reaches and warms the Earth. For this purpose, sulphur dioxide and other sulphur compounds are being tested for injection as aerosol. But this is likely to increase acid deposition on the ground and add to ozone layer depletion. At present, SRM is a theoretical tool and is fraught with high risk.

Experts believe that even if cumulative carbon dioxide emissions between 2000 and 2050 are restricted to 1000 billion tons, there will still be 25% chance of warming exceeding 2 °C (Meinshausen et al. 2009). At present, the cumulative CO_2 emission is around 36 billion tons annually with China, India, USA and Russia being the principal polluters (Fig. 5.1).

Large emitters like USA have committed to reduce their carbon emission by 26–28% from 2005 levels by 2030 meaning thereby that the US emission would be

5.1 Introduction

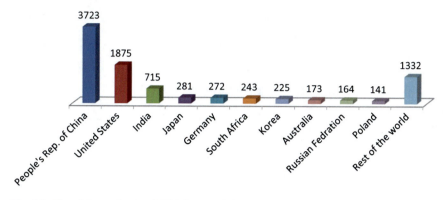

Fig. 5.2 Electricity production (TWh) from coal. *Source* IEA (2013)

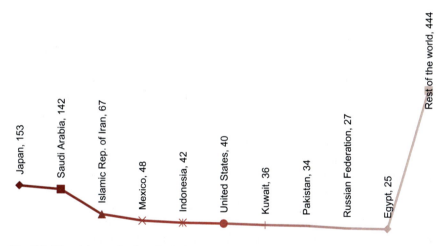

Fig. 5.3 Electricity production (TWh) from oil. *Source* IEA (2013)

4500 million tons in 2030 (Taylor and Branigan 2014). Thermal power plants in the USA contribute nearly a third of GHGs, and therefore, it will have to switch from coal to gas/renewable by 2030. China whose present CO_2 emission is 9000 million tons is expected to reach its peak of 18,000–20,000 million tons by 2030 (Li et al. 2016). India, the third largest emitter, contributes 2000 million tons of CO_2 at present and is expected to reach about 4000–5000 million tons by 2030. USA, China and India will have to necessarily reduce their dependence on coal in the coming decades.

What actions are required by individual nations between now and 2030 are a big question. Every nation, especially the middle- and poor-income countries, is more worried about their employment and economic growth than investment in energy-efficient technologies and innovations. This in turn means phasing out use of coal- and oil-based energy production (Figs. 5.2 and 5.3) and investing in low carbon/zero carbon system for industry, buildings and transport. Addressing the galloping and wasteful use of energy by rich and upper middle class is by no means a humongous effort. It just requires compromises in aspirations, consumption pattern and lifestyles.

5.2 Can We Postpone 'Energy Apocalypse'?

Rock, stone, dust is this earth;
This earth is supported, held together;
To this golden breasted earth I have rendered obeisance
What, O earth, I dig out of thee,
Quickly shall that grow again
May I not, O pure one,
Pierce that vital spot, (and) not thy heart

-*Atharva Veda* (1500–1000 BC)

The International Energy Agency estimates that in order to meet the energy efficiency target for 2050, the global spending has to increase from the current level of 300 billion USD per year to 680 billion USD, a seemingly impossible task. Even the energy demand of transport sector that consumes 27% of global energy will need to be revolutionalised. Most of the old vehicles consume very high quantity of fuel and are required to be changed dramatically in a time-bound manner, a non-feasible option for all developing countries that still use archaic engines of the 1990s or before. Similarly, buildings that consume nearly 30% of energy for lighting, heating, cooling and ventilation will be required to be converted to 'Net Zero' buildings, another gigantic task to achieve.

The global population is expected to cross 9 billion by 2050, and meeting the basic needs of food, shelter, livelihood, water and energy will be a major challenge that will be exacerbated by the uneven consumption pattern and waste generation among the high-income, middle-income and low-income population groups. Developing and low-income countries, particularly in Asia and Sub-Saharan Africa, currently need affordable energy at all costs to facilitate survival and growth strategy for more than 1.6 billion people who have no access to electricity and around 2.6 billion without clean cooking fuel. The number of people without electricity and gas has actually risen since 1990. While high- and middle-income groups have access to and can afford energy services from coal, oil, gas, nuclear, wind and other sources, most energy services for low-income group come from the biomass sources including firewood, charcoal, animal dung, crop residues and other combustible wastes. In a few countries like Chad, Malawi, Ethiopia, Tanzania and Somalia, 90% of the national energy consumption comes from wood. With the depletion of fuel-wood supply over the last several decades, more than 1000 million poor are forced to switch over to other cheaper/free substitutes such as straw, crop residues and animal dung. The burning of wood, dung and crop residues not only increases GHGs but also reduces mulch and fertilizer, thereby lowering agricultural productivity. As resources deplete, collection time increases, exacerbating the human health effects and also displacing

5.2 Can We Postpone 'Energy Apocalypse'?

time that would otherwise be spent on production (GDP) and/or leisure (health and happiness).

Energy, whether obtained from the primary sources (such as Sun and biomass) or secondary sources (such as electricity and hydrogen), is vital as it contributes significantly to GDP growth at the early and intermediate stages of development. Everything that we use or consume including the food, cloth, road, transport and building requires energy. A study of economic growth in selected developing countries shows that energy availability and consumption are still significant at the early stages of developments (e.g. up to about 20% in China and India). It has also been realized that countries with low scores on UNDP Human Development Index have low primary energy consumption per capita. In countries where more than 75% of the population receives less than $2 per day income, energy consumption (biomass plus commercial) is around 0.4 tons of oil equivalent (toe) per capita per annum. As the percentage of poverty indicator falls to 40–75%, energy consumption rises to about 0.8 tons of oil equivalent (toe) per capita per annum, and at 5–40%, it rises to over 1.5 tons of oil equivalent (toe) per capita per annum. Very few countries with less than 2 tons of oil equivalent per capita per annum achieve a high Human Development Index score.

A comparison of energy use per capita and GDP of selected few countries (Fig. 5.4) in the chart above shows that under the given set of circumstances when countries continue to use non-renewable and climate intimidating energy, Japan, China, India, Indonesia and USA are able to utilize their energy input efficiently whereas Australia, Canada, France, Russia, South Africa and others are inefficient performers in terms of GDP. These nations generate more than 90% of their electricity from non-renewable and contribute substantially to insidious GHG emissions without improving their GDP.

Whether correct or not, the fact remains that the cumulative effect of all such actions has made the world so disabled that even if we stop all fossil fuel burning it will take several thousand years to restore the normal health of this spaceship. Reversing this trend is going to be a formidable dilemma (between economic growth, energy security and climate change) as most of the low- and middle-income countries will strive for higher income thereby increasing energy demand to extend services to more than 1.6 billion people without electricity and 2.6 billion without clean cooking fuels.

Most of the countries are currently facing three energy-related threats:

i. Inadequate and insecure supplies of energy;
ii. Poor energy efficiency; and
iii. Environmental harm caused by overconsumption of energy (Fig. 5.5).

With the global economy expected to grow four times by 2050, an investment of $45 trillion will be required to bring in viable replacement for oil, gas and coal (The New Climate Economy 2014). Overhauling all existing electricity generating plants, that produce 2 million megawatts, and meeting additional demand of 6 million megawatts by 2030 will itself cost $5.2 trillion. The question then is whether countries can afford to share this insurmountable cost.

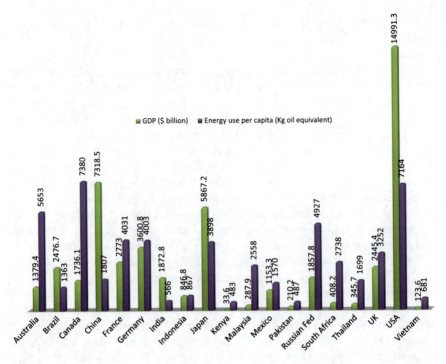

Fig. 5.4 A comparison of GDP and per capita energy use in a few selected countries. *Source* World Bank (2013)

A business-as-usual scenario suggests that energy efficiency will reduce energy demand by only 20% in developed countries by 2050 (Frei et al. 2013). This would essentially mean that the USA and European Union would cut oil and coal consumption significantly and substitute them with renewable resources. However, China, the largest producer and consumer of coal (70% of its commercial energy comes from coal), will face many challenges in reducing the share of coal in primary mix. India too relies heavily on coal (53% of its commercial energy) and will not be in a position to transit from non-renewable to renewable source in near future. At present, India relies heavily on biomass-based fuel (in rural and forest areas) which will further add to the carbon emission. India and China will, therefore, require huge upfront investment in energy sector in the coming 10 years to ensure that global consumption of fossil fuel is reduced from the current level of 80–50% by 2050. This will, in turn, require scaling up nuclear and renewable from the current level of 13–40% by 2050. Use of nuclear energy is expected to increase mainly in India and China and renewable mainly in North America and the European Union. In real terms, this will amount to setting up of 32 nuclear plants (1000 MW each), 17,000 wind turbines (4 MW each), 215 million square metres of solar photovoltaic panels and 80 concentrated solar power plants (250 MW each) over the next 35–40 years.

5.2 Can We Postpone 'Energy Apocalypse'?

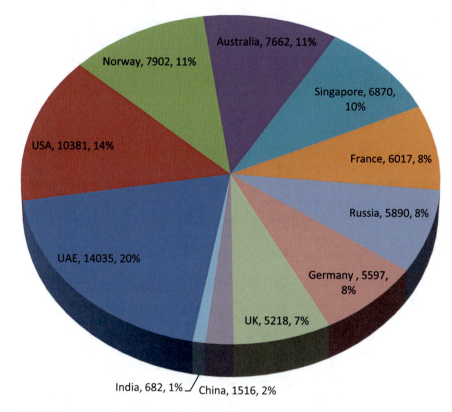

Fig. 5.5 Average energy use per capita in watts. *Source* Kalam and Singh (2011)

It is time for all developing and low-income countries to understand that production and supplies should not be restricted to the conventional sources and systems, viz. coal, oil, nuclear, thermal, hydroelectric and wood fuel alone. Other viable sources like wind, solar and biomass need to be considered for future options (Table 5.1). Energy-saving measures including efficiency, carrier substitution and alternative modes of energy usage should be treated as equivalent of energy supply enhancement. A rational approach is required for arriving at an energy basket involving centralized, decentralized and energy-saving measures.

The following message is significant for all those directly or indirectly connected to the conventional system of energy:

i. Energy is and will be required at every stage, be it plant, animal and human growth, physical movement, food, household use communication and so on. In other worlds, survival and development of humanity are directly proportional to the availability of energy.
ii. Transition from low- to middle-income country and middle- to high-income country will lead to high energy demand.

Table 5.1 Energizing economy with compromises (2050 scenario)

Suggestion	Possibility
Energy efficiency by almost all nations	Possible
Shift to renewable energy by almost all nations	Difficult
Low carbon intensity[a] energy by almost all nations	Difficult
Cost reduction in production and transport by almost all nations	Possible
Remove energy-related subsidies by almost all nations	Difficult
Save energy consumption by increasing cost by almost all nations	Difficult
Provide financial incentives for renewable (by almost all nations)	Possible
Impose carbon tax in all nations	Difficult
Provide clean cooking fuel to poor in almost all poor and developing nations	Difficult
Optimize/reduce energy consumption with increasing income in rich countries	Difficult
Increase fossil fuel prices to make renewable more competitive	Difficult
Conversion of oil-based transport sector to renewable energy in all developed countries	Possible

[a] Carbon intensity = units of CO_2 produced by a unit of energy consumed

   iii. That while fossil fuel reserves are depleting rapidly, their supply and consumption (especially coal, oil and gas) are increasing.
   iv. That enhancing economic growth and sustaining it without energy access, affordability and security to poor will be counterproductive.
   v. There is overwhelming scientific evidence that anthropological emissions of greenhouse gases from carbon combustion threaten catastrophic consequences from rapid climate change.
   vi. There are severe health and environmental consequences from fossil fuel combustion being experienced in every major developing city and town.
   vii. That substantial funds will be required for technological advancement in order to ensure that on the one hand renewable energy is available at affordable cost to all and especially poor, and on the other hand, the conversion ratio between products/services and energy input is very high.
   viii. That investment in both the quantity and quality of energy for the poor will substantially improve the chances of poverty reduction and GDP growth.

5.3 The Future

5.3.1 The Solar Energy

The solar energy directly absorbed by the atmosphere and the oceans primarily energizes water and air. Winds, clouds, air, cyclone, rain and snow are the manifestation of this absorbed energy that govern the physical, chemical and biological processes including converting the carbon dioxide from the air and water from soil or sea into energy-rich carbohydrates and oxygen. The oxygen so generated supplements the oxygen that was produced by primitive bacteria of ocean some 2 billion years ago, to permit living beings on the Earth to develop and evolve.

Besides providing direct energy, sunlight can also be transformed into electricity either by generating steam or through photovoltaic or solar cells. Solar cells absorb photons and convert them into electrons which are, in turn, collected by electrodes to flow into a well-defined circuit. The traditional solar technology uses silicon as semiconductor material. Steam is generated by focusing sunlight on a receiver by parabolic troughs or a series of flat mirrors called heliostats. The solar photovoltaic panels, made of silicon, are expensive and provide efficiency of 10–20%, but their discovery has revolutionized the energy industry. The silicon semiconductors absorb energy from the photons of shorter wavelengths only, and the energy from long wavelengths is lost. As a result, the current silicon solar cells convert only a quarter of the light that falls on them. Research is currently on to fix this problem by using a different material for each layer of the semiconductor (such as arsenic, gallium and indium). The objective is to use maximum light including infrared and UV which are otherwise invisible to human eyes. Each of these layers absorbs photons from different wavelengths, converts it into electrical energy and passes on the rest to the other layers. By using compounds like gallium indium phosphide and gallium indium arsenide and using a lens to concentrate sunlight, scientists have built photovoltaic cells that are 40.8% efficient. In future, it will be possible to produce relatively cheap and thin films of solar cells that absorb non-visible wavelengths of light and convert it into electricity.

Solar panels are used in a wide range of products starting from wristwatch to calculators and satellites. Newly constructed residential and office buildings in Japan, USA and Germany have incorporated photovoltaic material to generate electricity. Solar energy is also used for generating electricity through steam turbines. The Ivanpah solar thermal power plant (Ivanpah Valley) in California, USA, is the largest of its kind in the world with 347,000 mirrors reflecting the rays of the Sun onto boilers to generate 377 MW power energizing 140,000 homes in Southern California.

With constantly reducing cost of solar cells (from $70 per watt of production in the 1970s to less than $3.50 per watt in 2001), it is argued that solar photovoltaic-based energy could replace oil by 2030.

5.3.2 The Wind Energy

Wind or air is simply a combination of gases within Earth's atmosphere that remain in motion due primarily to uneven heating of the Earth's surface by solar energy. The kinetic energy of general circulation of the atmosphere is said to be equivalent to the energy released by the detonation of a powerful hydrogen bomb. Similarly, an average tropical thunderstorm generates energy that is equivalent to the detonation

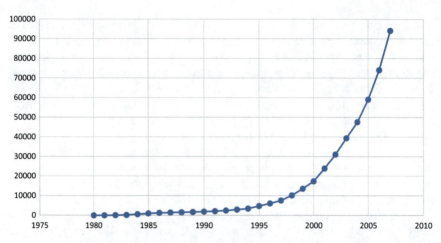

Fig. 5.6 World wind energy generating capacity (MW). *Source* Earth Policy Institute (2008a, b)

of a 20 kiloton atomic bomb. Wind energy is a vast and unending source that has the potential for substituting coal and oil in near future. And unlike hydropower and coal, it does not divert land use or interfere with it. For example, windmills can be conveniently set up in farmlands with least interference with the farming activities and capacity to transform rural economy (*1 MW of windmill generated electricity can supply 350 homes*) by enhancing their income and providing jobs.

The wind turbines are usually three-bladed structures with 10–22 revolutions per min that spins a shaft which is connected to a generator to produce electricity. Experiments are currently underway to use carbon and other composite materials instead of steel and glass.

The phenomenon of constant flow of wind can be effectively utilized for harnessing this clean and reliable source of energy especially in areas having strong wind velocity and long spells. Once harnessed, it can be used for applications such as water pumping, battery charging and power generation.

Wind energy is considered to be the fastest growing user-friendly energy source (Fig. 5.6). The Dutch started exploiting wind energy way back in the seventeenth century and currently meets 5.2% of their energy needs from wind technology (The Economist 2015c). To upscale its renewable energy generation to 14% as mandated by EU, the Netherlands is planning to have offshore wind parks in the North Sea with a capacity of 3450 MW (Green and Vasilakos 2011) (Fig. 5.7).

Outside EU, countries like China, USA and India (Fig. 5.8) are taking big leaps in improving their installed generating capacity from wind. In China, wind power has tripled its share since 2010 and is capable of lighting 110 million households (Ren21, Renewables 2016).

USA is also venturing into offshore wind farms by installing turbines for the residents of Block Island, a small island off the coast of Rhode Island. It is a pilot pro-

5.3 The Future

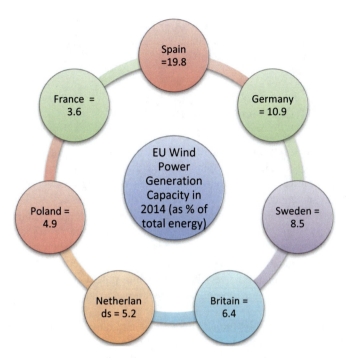

Fig. 5.7 EU wind power generation capacity in 2014 (as % of total energy)

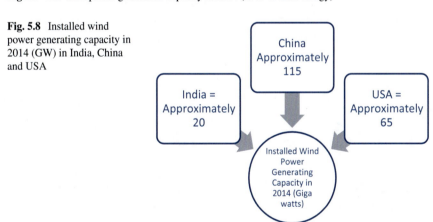

Fig. 5.8 Installed wind power generating capacity in 2014 (GW) in India, China and USA

gramme, and if successful, several thousand offshore wind turbines will be installed to light up an area equivalent to New Jersey.

With the cost of wind power generation dropping from USD 2600 in 1981 to USD 800 in 1998, wind power production started gaining momentum from its naissance in California in the 1980s. In the decade of the 1990s, it spread to countries like Denmark, Germany, Spain, USA, India, Netherlands, UK and China expanding phenomenally between 7500 MW in 1997 and 15,000 MW in 2000. In 2005, world's total installed wind energy capacity was 59,091 MW, and by 2006, it reached 74,223 MW showing a record growth of 41%.

5.3.3 The Hydro-/Geothermal Energy

The burning of Sun, besides providing direct energy, sets in motion all the biogeochemical cycles on the Earth. Of the 3% freshwater, only a minuscule (0.03%) is accessible for human use through the hydrological recycling process of evaporation, transpiration and condensation. An appreciable part of this freshwater that is lodged in snowcaps, retained in surface soils, lodged in natural aquifers and drained through streams and rivers into the oceans becomes a major source of energy directly (through consumption of water and aquatic food) and indirectly (through irrigation, industries and electricity generation). Flowing water in the river systems, captured in man-made reservoirs (or run of the river), has tremendous potential to generate energy for human use and consumption through small, medium and large hydroelectric projects (Fig. 5.9).

Small hydropower unit is probably the oldest and most reliable of all renewable sources where energy in the form of electrical power is generated through the use of falling water under gravitational force. Over the years, small hydropower plants have been made attractive through standardization, better planning and management, to reduce initial cost and commissioning schedules. Many of these projects are an attractive renewable energy source of decentralized power in remote and hilly regions

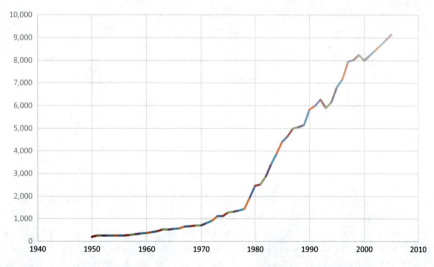

Fig. 5.9 Cumulative growth of geothermal installed capacity (MW). *Source* Earth Policy Institute (2008b)

5.3 The Future

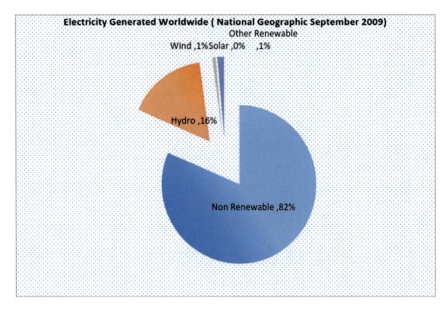

Fig. 5.10 Electricity generated worldwide. *Source* Johnson (2009)

as they can be easily set up along small streams, medium to small rivers, with sufficient water flows. The greatest advantage of such projects is that they do not cause deforestation, submergence or rehabilitation. Such projects are able to accelerate economic development in protected landscapes with greatest benefits.

Though small hydel units have distinct socio-economic, technical, environmental and operational advantages, most of the present-day hydropower generation is through large projects, which require high initial investment, submergence of extensive forest tracts, negative environmental impacts and other externalities that lead to long gestation period. China, the world leader in large dams, has constructed more than a thousand large dams since 1949, the largest being Three Gorges with a capacity of 22.5 GW. Currently, hydropower is produced in more than 150 countries and accounts for nearly 16% of global electricity generation producing 3427 TWh (Fig. 5.10).

Besides hydel generation, humans have also made use of other inexhaustible sources of energy such as hot springs, geysers and volcanic activities for centuries (hot bath and steam). The first reported use of geothermal energy for commercial electricity generation was by Italy in 1904. During the twentieth century, the cumulative geothermal energy generating capacity reached 1100 MW in 1973, and by 1998, it grew eightfold to 8240 MW, thanks to oil crisis of 1973. USA has the world's highest installed capacity of geothermal energy generation at 3.4 GW. Technologists are now working on enhanced geothermal systems whereby millions of gallons of water, sand and chemicals are injected into vertical wells at relatively high pressure

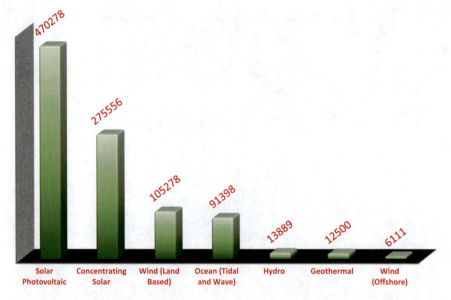

Fig. 5.11 Potential for global electricity generation from renewable. *Source* (in TWh) Johnson (2009)

to create fresh fractures for creating a network so that the pumped-in cold water is heated and sent back to the Earth surface for generating power.

It is estimated that 40% of the world's geothermal energy resources are located in Indonesia (equivalent to 28 GW or 12 billion barrels of oil). Currently, around twenty countries (on the Pacific Rim, bordering the Mediterranean Sea and along Africa's Great Rift) have tremendous potential to tap the geothermal energy and to produce enough electricity to meet the substantial requirement in Japan and Indonesia (Fig. 5.11).

5.3.4 Energy for Batteries (Lithium: The New Gasoline)

Almost everyone today uses batteries for staying connected either through phones, computers, vehicles or by any other means. Trillions of mobile phones and electric cars are powered by lithium-ion batteries commercialized in the 1990s (Singh and Singha 2016). The future of world automobile will soon witness a revolution with the introduction of lithium batteries. Lithium (Table 5.2) is the lightest metal on the planet that can hold charge extremely efficiently as compared to heavier batteries of the past made of lead, zinc and nickel–cadmium. Mostly available in nature as a viscous liquid, lithium chloride is converted into finely powdered lithium carbonate and hydroxide which is then used commercially forming a vital component of batteries that energize everything from cars to smartphones, computers and power tools. It is excellent

Table 5.2 Total lithium reserves = 40 million tons

Country	Identified lithium resources (% of total)
Bolivia	22.7
Chile	18.9
Argentina	16.4
USA	13.9
China	13.6
Australia	4.3
Others	10.2

Source Borthomieu (2014)

at holding a charge and delivering electric current to the vehicle without frequent recharging stops. One of the major drawbacks of lithium is that it is highly reactive and overcharging or manufacturing defects can lead to short circuiting causing the batteries to heat up and burst into flames occasionally.

The lithium-ion battery has four components, viz. copper, lithium cobalt oxide, plastic and graphite where lithium is positive electrode and graphite is negative (Mikolajczak et al. 2012). On an average, a lithium battery can store 100–250 W-h/kg more than twice that of nickel–cadmium. A small electric car with 25 kW-h lithium battery can cover 175 km before recharge.

Though the current sale of lithium salt is only 1 billion USD a year, the prices have risen sharply from USD 6 per ton in 2015 to USD 14 in 2016. Lithium batteries are relatively expensive costing USD 500/kW-h capacity (Berckmans et al. 2017). This means that a battery pack for a small electric car would cost USD 10,000 which is pretty expensive for a common man to switch over to the new cars. Research is currently on to improve the energy density of lithium batteries by developing thin films (single atom thick) to enclose positive electrode coupled with a negative electrode of sulphur (like lithium, sulphur also has a very high energy capacity). This will enable a battery to hold about five times as much energy by weight as compared to lithium batteries.

China has tripled its production and sale of electric cars between 2014 and 2015. Similarly, Tesla Motors, an American electric car manufacturer, is planning to produce 500,000 electric cars in the next 5 years using magical element lithium. Besides, car batteries and lithium batteries are being used for storing power generated through solar panels and windmills in the USA, Australia and South Africa.

5.3.5 The Biomass Energy

All green plants containing chlorophyll whether in the forests or outside in the form of agriculture crop or otherwise absorb solar energy and carbon dioxide to produce biomass for growth and reproduction. It is interesting to note that only 2% of solar energy is absorbed by plants, and this energy is sufficient to feed billions of humans and trillions (rather zillions) of animals on Earth.

Biomass or standing crop = Gross Primary Productivity of a given area—respiration by plants, animals and saprophytes

Gross Primary Productivity − Respiration(carbon dioxide is released by plants)
= Net Primary Productivity
Gross Primary Productivity = Solar Energy + Plants

Since respiration is a continuous process, Gross Primary Productivity (of a plant or all plants in a given area) is a dynamic process whereby plants accumulate biomass and use part of the biomass for respiration. As a general principle, most of the energy built up during Gross Primary Productivity is utilized for metabolic activities and only 20% of the GPP energy goes in the standing crop of trees, plants and animals.

The total growing stock of the global forest, in the year 2005, was estimated at around 434 billion cubic metre that stored 283 gigaton of carbon in their biomass and 638 gigaton of carbon in the terrestrial ecosystem as a whole (up to a soil depth of 30 cm).

Early man started using biomass (leaf, twig, root and stem) as a major source of energy for fulfilling the daily needs mainly for food and warming the houses in colder regions. From then on, biomass has remained an indispensable source of domestic energy for billions of humans particularly in developing nations. Over half of the biomass using people live in China, India and Indonesia, and the reasons for this heavy reliance are not difficult to understand. Firstly, the supply of biomass is assured vis-a-vis energy from thermal, nuclear, oil, hydro and other sources. Secondly, the cost, if any, is within the paying capacity of the local population, and lastly, the sociocultural factors favour the use of biomass fuel for cooking food rather than using coal or gas. Currently, biomass energy accounts for nearly 5% of the world's total energy supply which ranges from 1% (in Europe) to 75% in Ethiopia and Congo. Seventy-two percentage of the total rural households and 33% of the total urban households in India use wood as fuel.

Non-wood biomass-based fuels or biofuels in short are renewable fuels which contain no petroleum, and can be blended at any level with petroleum fuel to create a biofuel blend. The emergence of non-wood biomass fuel started with the energy crisis of the 1970s that led many countries to seek ways of improving their energy security by decreasing their dependence on fossil fuels and diversifying their energy supplies. While biofuel production has never really been significant due to the low price of oil, the role of biomass as a fossil fuel energy substitute has regained a great deal of interest due to:

- *Instability in petroleum-producing countries;*
- *The rising cost of petroleum from less than US$ 20 per barrel in 1995 to more than US$ 60 in 2006 and over US$ 100 in 2013;*
- *The adoption and entry into force of the Kyoto Protocol, which requires ratifying countries to reduce GHG emissions;*

5.3 The Future

- Known petroleum reserves are estimated to be depleted in less than 50 years at the present rate of consumption;
- Realization that biofuel production can bring energy security, protect from energy pricing risks and result in significant savings in foreign exchange.

Currently, there are two widely recognized biofuels for the transportation industry:

Bioethanol, *which is produced out of plant starch, sugar, and more recently, but still on an experimental basis, cellulose; and*

Biodiesel, *or* diesel produced from vegetative sources, was first used by Rudolf Diesel in 1893. However, it did not attract the attention of commercial nature due to abundance of oil, gas and coal till the second half of the twentieth century when the issue of global warming attracted global attention. It has now been estimated that biodiesel produces 78.5% less CO_2 emission as compared to burning of petroleum diesel. Biodiesel *is made out of vegetables or grain oil and recycled cooking oil. It is cheap and renewable, biodegradable, non-toxic, is safe to store (due to a high flash point), is kinder on the engines, has a long shelf life, is free of sulphur and aromatics, releases no more carbon dioxide than the trees originally consumed and produces significantly lower emissions.*

Both bioethanol and biodiesel can be blended with gasoline or petroleum-based diesel or even pure in flex-fuel cars. The major biomass sources currently used are sugar cane and corn to produce bioethanol and rapeseed and palm oil for biodiesel production. Other sources such as sunflower seeds, soybean, peanuts, jatropha, castor bean, corn and coconut oil are also used for biodiesel, and wheat, sugar beet, sweet sorghum and cassava for bioethanol.

Biofuels account for 23% of all road transports in Brazil, 5% in the USA and 4% in the European Union. In 2011, 99% of biofuels produced and consumed were made from food crops and 95% of ethanol produced by USA is consumed in the country (Gerbens-Leenes et al. 2012).

Brazil took lead in biofuel production from sugarcane after the first oil crisis of 1973–74, as an alternative to reduce their reliance on crude imports. Currently, it produces 15 of the 25 billion litres of ethanol used worldwide for energy production. It has also developed local innovations such as **'flex-fuel'** cars allowing drivers to choose fuel depending on its price. Promoting biofuel as a partial substitute for fossil fuel is a prudent approach being followed by several countries. Fuel alcohol, biodiesel, hydrogen and biofuel cell are some of the current interests, and several technological improvements have been made to produce biofuel in an economically viable manner. While developed countries use edible oils for biodiesel production, developing countries have opted for cost-effective non-edible oils for their production. There is a huge potential for biodiesel growth in Europe and America due to its environmentally friendly properties.

South African Airways in collaboration with Boeing is currently engaged in developing biofuel from seeds of specially bred strains of the tobacco plant (Powell 2015). The airline is expected to use 20 million litres of fuel so produced in equal proportion with conventional jet fuel by 2017 (Table 5.3).

Table 5.3 Characteristics of liquid biofuel

Biological species	Fuel	Fuel yields (L/ha)
Corn grain	Ethanol	3000
Sugar cane	Ethanol	6000
Sugar beet	Ethanol	5000
Wheat	Ethanol	2500
Soya bean	Diesel	500
Rapeseed	Diesel	1100
Sunflower	Diesel	1000
Palm oil	Diesel	4500

5.3.6 The Biogas Energy

With the declining supply and externalities associated with coal and petroleum, many countries are opting for biogas as a climate-friendly substitute. Biogas generated by anaerobic fermentation of organic waste contains 60% methane, 40% carbon dioxide and up to 20 ppm hydrogen sulphide. Under high pressure, carbon dioxide separates out enhancing the calorific value of the remaining gas. It is also possible to obtain 90% pure methane at very high pressure and low temperature as carbon dioxide solidifies at a pressure of 46 atm (673 psi) and at minus 68 °C. This methane can be either supplied directly or bottled and transported as a substitute for petroleum fuel. It is estimated that more than six tons of biogas can be generated per day from municipal waste generated by a population of four lakh. Biogas from waste is extremely useful for villages and small towns of countries like India where unmanaged waste is a health hazard and people do not have paying capacity for commercial electricity.

5.3.7 The Hydrogen

Hydrogen is being experimented as a fuel for transportation. Compared with electric batteries, the hydrogen fuel cells work by consuming hydrogen that reacts with oxygen from the atmosphere (platinum nanoparticles acting as a catalyst) to produce water and electricity. At present, hydrogen fuel cells are running at 60% efficiency. The reason is that platinum, the catalyst, is dispersed by high surface area carbon support that has a tendency to corrode while switching on and off the engine. Researchers are working on alternatives like titanium ruthenium oxide to support platinum nanoparticles. Besides, solar technologies also have the potential to generate electricity by electrolyzing water and splitting it into hydrogen and oxygen and then using hydrogen-powered generators to produce fuels for automobiles. Since hydrogen can be stored, it complements energy produced with wind and solar power.

5.3 The Future

5.3.8 Energy from Waste

In today's world of technology, nothing is worthless not even waste that we produce. As the organic waste decomposes, the aerobic bacteria remove oxygen and the anaerobic bacteria react with the waste to produce acetate which is then converted into carbon dioxide and methane. The USA is one of the few countries where food is wasted in huge quantity. In 2013, the cumulative organic waste was 250 million tons. More than 600 landfill projects are currently in operation covering almost all states generating 15 billion kilowatt-hours enough to power roughly 1 million homes for a year.

Another area of generating power is by using the property of some materials whereby heating part of an object made of that material drives electrons from hot part to cold part creating a current (Seebeck effect). Tapping this heat, converting it directly into electricity is being explored by scientists. Strontium titanium oxide is one such material that can be used when heated between 700 and 750 °C.

Heat generated from heavy machineries such as boilers, diesel engines, thermal power plants normally goes waste and increases the temperature of the surrounding environment. A new technology called Organic Rankine Cycle can convert this waste heat from all kinds of heavy machinery into electricity. This technology can be easily integrated into the existing industrial process.

5.4 On Way Out

5.4.1 Coal

The power of sunlight captured millions of years ago by plants and animals and buried in huge deposits is now being burned as coal, petroleum and natural gas. The earliest reference to the use of coal as fuel by humans is from the geological treatise *'On Stones'* by Theophrastus where he describes different marbles and mentions two types of coal, one which retain their heat and can be rekindled by fanning and the other that was ignited by the heat of the Sun especially when sprinkled with water. Englishmen started using coal for fuel as early as ninth century which was subsequently banned during the reign of King Edward I (1272–1307). Richard II (1377–1399) revoked the ban and introduced taxation. This was succeeded by strict regulatory measures for use of coal by Henry V (1413–1422). Unfortunately, by the sixteenth century, much of the natural forests in England were cleared to meet the growing requirement of fuel, timber, shipbuilding and farmland. Englishmen were left with no option but to use coal as a substitute for wood energy.

With the onset of the Industrial Revolution in the eighteenth century, the consumption of coal shot up rapidly. Coal replaced wood as a major source of fuel for transportation, industries, household energy and for electricity generation earning the reputation of being the workhorse of the world energy economy. It continues to be responsible for 41% of the global electricity needs and is the second best source of primary energy after oil and gas and tops the list of sources of electricity generation. As concerns about urban air pollution and global climate change escalated in the closing decades of the twentieth century, technical experts realized that climate change was real and its impact might cause massive destruction if adaptive strategies were not adopted by most of the countries of the world. As a result, with the beginning of this century, the Sun began to set on the fossil fuel era. It is the desire for clean climate-benign fuels on the one hand and the depletion of fossil fuels on the other that is driving the global transition to the non-conventional energy era. Coupled with this is the sharp increases in the prices of coal, oil, natural gas and uranium that underline the importance of alternative energy sources and innovative technology for all countries. At present, oil, gas and coal are responsible for meeting 70–75% of global energy requirement most of which goes into generating electricity and transportation. Hydroelectricity meets 5%, nuclear 6% and biomass around 10–12%. Only 1% energy comes from solar, wind, tidal and geothermal sources. Coal is responsible for meeting 41% of the world's electricity and 29% of the world's energy needs. Coal production has been on the rise in China since 1995. From nearly 3 billion tons per annum in 1995, China expanded its production to 3.9 billion tons of coal in 2014 to fuel its rising economic demand. In order to maintain GDP at 8% per annum, China will have to generate more than 9200 TWh in 2030 and the current trends indicate that it has no other options but to burn coal to supply 70% of its energy needs. If the trends continue, China will continue to lead the world as the largest emitter of CO_2 and other global warming gases. Reducing current levels of GHG and sequestering the existing ones is an uphill task which indicates a massive risk to all living beings, infrastructure as well as continued melting of Himalayan glaciers.

India too is all set to increase its coal production to 550 million tons during 2015–16.

5.4.2 The Hydrocarbons

All hydrocarbons whether oil or gas are generated in the sedimentary rocks from the organic matter of dead animals and plants that are deposited along with the sediments, and the physiochemical changes of thermogenic and biogenic origin converts the organic matter into oil/gas. Temperature ranging from 50 to 150 °C deep inside the Earth (geothermal gradient) is conducive for generation of oil and gas, and the whole process is called catagenesis. This temperature range is called the 'oil window', and it signifies the depth where the process of conversion of organic matter to hydrocarbons is initiated and completed. The depth of oil window, nevertheless, varies from place to place as the geothermal gradient varies from place to place.

The quality of crude oil varies from place to place ranging from watery to almost solid in viscosity, fawn to deep black in colour and from low to very high sulphur content. Globally, more than hundred grades are traded of which West Texas Inter-

5.4 On Way Out

mediate and Brent are the benchmarks. A barrel of Brent was $109 in 2013, and WTI (West Texas Intermediate) was $98.

After petroleum, natural gas is the most widely used energy source in the world overtaking oil consumption in the year 1999 and increasing its use by 12 times during the second half of the twentieth century. Natural gas was first sold internationally in the 1960s as a substitute for oil in heating and power generation. It is almost a clean source of energy from environmental angle as it emits less than half carbon as compared to coal. Global export of natural gas is expected to rise from 290 million tons per year in 2013 to 400 million tons per year in 2018 (Fig. 5.12).

5.5 The Debate on Nuclear Energy

Energy generated by the nucleus of a fissile material is undeniably far more powerful as compared to the one generated by conventional fuel. For example, the energy generated by 10,000 tons of coal can be produced by 500 kg of naturally occurring uranium or 62.5 kg of naturally occurring thorium (500 kg of naturally occurring uranium normally contains 3.5 kg of fissile uranium-235 fuel). Currently, there are 29 countries operating 441 nuclear plants (reactors) generating about 375 GW of electricity. A nuclear reactor is a device containing fissile material in sufficient quantity and arranged in such a manner that facilitates a controlled nuclear fission reaction. The reactor breaks up the nuclei of fissile material such as uranium-235, uranium-

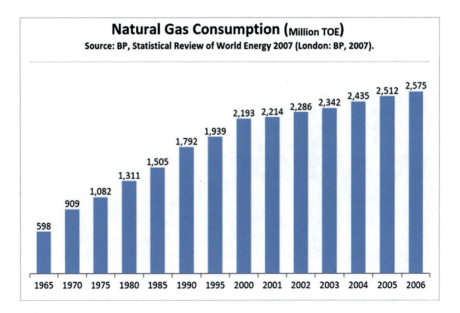

Fig. 5.12 Natural gas consumption (MTOE)

Table 5.4 Share of nuclear power

Country	Ranking in the order of total nuclear power	Total nuclear power capacity (MW)	Share of nuclear power in total energy generated (%)	GDP (PPP) per capita
USA	1	101,433	19.59	46,860
France	2	63,130	74.12	33,910
Japan	3	47,348	29.21	33,885
Russia	4	23,084	17.09	15,612
Republic of Korea	5	18,785	32.18	29,997
Ukraine	6	13,168	48.18	6698
Canada	7	12,044	15.07	39,171
UK	8	10,745	15.66	35,059
India	15	4385	2.85	3408

Source Kalam and Singh (2011)

233 or plutonium-239. The fission reaction is accompanied by a slight loss of mass of the original material which manifests itself in the form of energy in accordance with Einstein equation $E = mc^2$. A typical reactor follows the following steps:

- Splitting of the uranium atom in the reactor core to generate heat;
- Pressurized water that acts as a coolant and carries heat to the steam generator; and
- Turbine that is driven by steam (Table 5.4).

Besides the conventional nuclear reactor, nuclear energy is also produced in breeder reactor. A conventional reactor normally uses enriched uranium (up to 3% of uranium-235) as fuel and either plain water, heavy water or graphite as moderator to slow down the speed of neutrons. A breeder reactor operates with high-speed neutrons, without any moderator, and converts uranium-238, available in the burnt up fuel of conventional reactors, into plutonium-239 or thorium into uranium-233. The plutonium-239 and uranium-233 can then be recycled as fuel in breeder reactors to generate still more plutonium and uranium-233. In short, a breeder reactor has twin advantages over the conventional one:

- It extracts much more latent energy in natural uranium.
- It can also convert thorium into useful nuclear fuel (U-233).

For example, one ton of natural uranium can provide energy equivalent to 10,000–20,000 tons of coal whereas the same amount of natural uranium can provide energy equivalent to three million tons of coal in a breeder reactor.

Most of the economically prosperous nations are currently using nuclear energy in a big way, and it forms a substantial proportion of their clean energy portfolio reducing the burden of climate change mitigation. Many lessons have been learnt

5.5 The Debate on Nuclear Energy

from the nuclear accidents, and accordingly, designs of nuclear reactors have been improved over last several decades to make them more tolerant of human error and reducing pollutants to almost zero level (except radiations). The fact, however, remains that the setting up of a nuclear reactor entails a huge cost that escalates with time overrun. For example, the cost of Long Island reactor in USA escalated from 300 million USD in 1973 to 5.5 billion USD in 1984 when it was finally completed. The delay was mainly attributed to public outcry and design changes. Besides, several billion dollars have been sunk in nuclear energy projects that were initiated but never commissioned due to public anxiety for reasons including accidents, both major and minor. Unfortunately, the issue of a permanent repository for nuclear waste has not been resolved as yet. The quantity of nuclear-spent fuel has increased enormously, much of that is being stored temporarily at the commercial power plant itself.

Unlike the nuclear bombs that are designed to deliver a huge amount of energy in a short time, civilian nuclear energy, on the other hand, delivers a small amount of energy over a larger time frame. Nevertheless, the use of nuclear energy comes with the baggage of risks and threats. And these are:

1. Safety track record of nuclear reactors of different countries;
2. Impact of accidental fallout and radiation on plants, animals and human beings;
3. Preparedness of nuclear nations to deal with the following:
   a. Enemy attack on nuclear installation;
   b. Theft of nuclear fuel including spent fuel;
   c. Earthquake, tsunami, cyclone and other natural disasters;
4. A comparative analysis of disasters (small as well as large) involving nuclear plants, thermal plants, hydropower plants and non-conventional plants all over the world;
5. What are the 'worst-case scenarios' and our preparedness to deal with them especially in terms of the scale of deaths, disorders and damage to property (natural and man-made) as well as recovery and restoration;
6. How soon the disaster-affected sites become habitable for humans, plants and animals? In other words, how long does it take to decontaminate trees, farms, water, animals, buildings, etc.;
7. What is the life of a nuclear power plant and how safe is the decommissioning process;
8. What is the assurance that the spent fuel dumped in deep sea or Earth will remain safe <u>forever</u> and is under no threat from natural and man-made disasters?

5.5.1 Environmental Effects of Nuclear Power Plants and Explosion

The spent nuclear fuel from nuclear fission plants using uranium and plutonium contains more than hundred carcinogenic wastes such as **strontium, krypton, ameri-**

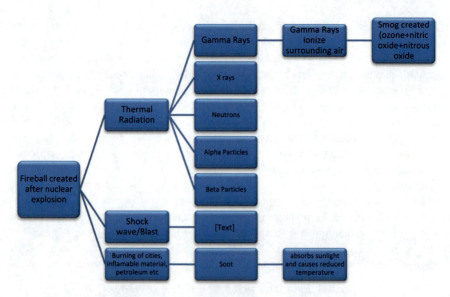

Fig. 5.13 Sequence of events after nuclear explosion

cium and cerium. Besides, the spent nuclear fuel and high-level reprocessing and plutonium wastes require well-designed storage for as long as thousands to a million years to minimize radioactivity spread into the environment. Safeguards are also required to ensure that neither plutonium nor highly enriched uranium is diverted to weapon use. There is general agreement that placing spent nuclear fuel in repositories' hundreds of metres below the surface would be safer than indefinite storage of spent fuel on the surface. Besides, many commercial nuclear power plants also release liquid and gaseous effluents that make civilians living within 80 km receive about 0.1 μ Sieverts per year (Fig. 5.13).

Radiation effects from nuclear explosions occur in three ways:

i. Exposure from radiation emitted by fireball which may be short duration but intense;
ii. Fallout from the ground burst as the particles begin to settle on the ground;
iii. Long-term exposure to low levels of radiations lasting years.

The normal annual radiation exposure include from 0.08 (medical X-rays) to 0.1 rem (natural radiation). However, in those areas natural radioisotopes are concentrated such as Kerala, Sri Lanka and Brazil; higher levels of exposures (0.5–12 rem) have been recorded. After any nuclear explosion, radio-nuclides like caesium 137 **(caesium 137 binds strongly to fine soil particles)** enter into the Earth and spread to near and distant locations through rain and wind (Table 5.5).

Nuclear Waste

During the cold war period when USA and USSR were fiercely competing for nuclear weapons, most of the spent fuel was disposed of without much precaution.

5.5 The Debate on Nuclear Energy

Table 5.5 Exposure effects of radiation

REM[a]	Exposure effect
Below 100 REM	Detectable changes in RBC, drops in lymphocytes, drop in platelet count, atrophy of lymph glands, loss of sperms, etc.
100–200 REM	Nausea, vomiting, death of blood cells, loss of appetite, fatigue, illness, infection, etc.
200–400 REM	Severe illness, nausea, vomiting, diarrhoea, haemorrhage (uncontrolled bleeding) of mouth
400–600 REM	Severity of above symptoms, increased chances of mortality
600–1000 REM	Bone marrow is completely destroyed; gastrointestinal tissues are severely affected ending in death from internal bleeding and haemorrhage

There are three units of radiation measurement: roentgen, rads and rems. A roentgen measures the amount of ionizing energy in the form of photons (gamma rays and X-rays) to which an organism is exposed. A rad is defined as the absorption of 100 ergs per gram of tissue and includes all forms of ionizing radiations that are actually absorbed by the organism. A rem is calculated by multiplying the dose in rads for each type of radiation and summing them

Liquid waste was buried in tanks from where it could leak into groundwater. Four hundred and thirty-six power generating nuclear reactors in 30 countries have accumulated thousands of tons of spent fuel and continue to add 10,000 tons a year (Mason 2013). The spent fuel (mainly uranium pellets) is normally encased in zirconium that is then put into the water pool for several years to absorb excess heat and some radiation. Thereafter, it is moved into huge concrete containers for permanent disposal in deep storage repositories.

5.5.2 Is Nuclear Energy Futuristic?

While there are several arguments against nuclear energy, the proponents argue that developing countries that face shortage of finance, infrastructure, efficiency and technology will take several decades to shift to safer options like solar, wind and hydrogen. The choice for such countries is limited in near future, either upscale consumption coal and petroleum and dump GHG gases in the atmosphere or shift to nuclear technology. China plans to triple its nuclear power generating capacity by 2020, and other emerging markets including India are also building new plants (The Economist 2015b).

Nonetheless, the trends in developed countries show that nuclear energy is being phased out for the following reasons:

a. Prohibitive cost of building a nuclear reactor;

b. Difficulty in finding storage sites for nuclear waste;
c. Subsidy in wind, solar and other renewable energy sources;
d. Reduction in price of natural gas making it difficult for nuclear power to run with economic efficiency;
e. Antinuclear wave after Fukushima.

The current status is that Germany, that once produced 20% of its power through nuclear plants, has shut down 8 of its 17 nuclear plants and will close all by 2022. France has decided to cut its share of nuclear energy generation from 75 to 50% in the next 10 years; Sweden has decided to shift from nuclear energy to wind farms; since Fukushima, all but two of Japan's 43 reactors have been suspended. Currently, only 394 nuclear plants are in operation, down 37 from 431 in the pre-Fukushima phase in 2010 (Brekke and Johansson-Stenman 2008).

US researchers are currently working on developing a nuclear fusion reactor that combines light atomic nuclei with heavier particles. And if successful, it may change the energy dynamics in the world in the sense that energy generation will have zero carbon emission. Scientists at the National Ignition Facility, Lawrence Livermore National Laboratory in California use a bank of 192 powerful lasers which are fired at a golden capsule that holds a 2 mm spherical pellet coated with fuel (hydrogen isotopes, viz. deuterium and tritium) on its inner surface. The laser beam (generating two megajoules of energy) hitting the gold capsule helps in emitting X-rays which in turn heat the pellet so high that it implodes with ferocious velocity resulting in the release of 17 kilojoules of energy (Table 5.6).

5.6 Innovations

The LED
After testing 1600 different materials including coconut fibre, fishing line and human hair, Thomas Edison finally zeroed on to a particular species of bamboo whose carbonized form was used as a filament in the first incandescent lamp. Researchers continued their efforts for improving the quality of lamp making the brighter and long lasting. Their efforts paid off at the beginning of the twentieth century with the introduction of filament made of tungsten that illuminated most of the households and streets for nearly a century. Meanwhile, in the 1960s light-emitting diodes (LEDs) first appeared in the market as indicators for electrical equipment and have been widely used in digital clocks, computer screens and traffic signals. These diodes used germanium and silicon and were more efficient in converting electricity into light as compared to the filament. LEDs are semiconductor devices that produce visible light when electric current passes through them. These are highly efficient sources of light in the sense that the illumination of 8 W LED is equivalent to 60 W incandescent bulb or a 60 W compact fluorescent lamp (CFL). Besides, the LED bulb has 20 times more longevity than incandescent bulb, produces less heat and does not contain mercury. According to US Department of Energy, by 2027 use of

5.6 Innovations

Table 5.6 Nuclear power plant accidents between 1952 and 2011

Date	Location	Description
Thursday, October 10, 1957	Sellafield, Cumberland, UK	A fire at the British atomic bomb project destroyed the core and released an estimated 750 terabecquerels (20,000 curies) of radioactive material into the environment
Tuesday, January 03, 1961	Idaho Falls, Idaho, USA	Explosion occurred at the National Reactor Testing Station, and three operators were killed
Wednesday, October 05, 1966	Frenchtown Charter Township, Michigan, USA	Partial core meltdown at the Enrico Fermi Nuclear Generating Station
Tuesday, January 21, 1969	Lucens reactor, Vaud, Switzerland	Loss-of-coolant accident, leading to a partial core meltdown and massive radioactive contamination of the cavern
Sunday, May 28, 1905	Sosnovyi Bor, Leningrad Oblast, Russia	Partial nuclear meltdown in Leningrad nuclear power plant reactor unit 1
Sunday, December 07, 1975	Greifswald, East Germany	Electrical error caused fire in the main trough that destroyed control lines and five main coolant pumps
Monday, January 05, 1976	Jaslovské Bohunice, Czechoslovakia	Malfunction during fuel replacement
Tuesday, February 22, 1977	Jaslovské Bohunice, Czechoslovakia	Severe corrosion of reactor and release of radioactivity into the plant area, necessitating total decommission
Wednesday, March 28, 1979	Three Mile Island, Pennsylvania, USA	Loss-of-coolant and partial core meltdown due to operator errors. There is a small release of radioactive gases. See also Three Mile Island accident health effects
Saturday, September 15, 1984	Athens, Alabama, USA	Safety violations, operator error and design problems force a 6-year outage at Browns Ferry Unit 2
Saturday, March 09, 1985	Athens, Alabama, USA	Instrumentation systems malfunction during startup, which led to suspension of operations at all three Browns Ferry Units
Friday, April 11, 1986	Plymouth, Massachusetts, USA	Recurring equipment problems force emergency shutdown of Boston Edison's Pilgrim Nuclear Power Plant
Saturday, April 26, 1986	Chernobyl, Ukrainian SSR	Steam explosion, fire and meltdown forcing evacuation of 300,000 people. Radioactive material dispersal across Europe

(continued)

Table 5.6 (continued)

Date	Location	Description
Sunday, May 04, 1986	Hamm-Uentrop, Germany	Experimental THTR-300 reactor releases small amounts of fission products to the surrounding area
Tuesday, March 31, 1987	Delta, Pennsylvania, USA	Cooling malfunctions and unexplained equipment problems
Saturday, December 19, 1987	Lycoming, New York, USA	Malfunctions force Niagara Mohawk Power Corporation to shut down Nine Mile Point Unit 1
Friday, March 17, 1989	Lusby, Maryland, USA	Cracks at pressurized heater sleeves, forcing shutdowns
Sunday, March 01, 1992	Sosnovyi Bor, Leningrad Oblast, Russia	An accident at the Sosnovy Bor nuclear plant leaked radioactive gases and iodine into the air through a ruptured fuel channel
Tuesday, February 20, 1996	Waterford, Connecticut, USA	Multiple equipment failure and leaking valve forces shutdown
Monday, September 02, 1996	Crystal River, Florida, USA	Balance-of-plant equipment malfunction forces shutdown and extensive repairs at Crystal River Unit 3
Thursday, September 30, 1999	Ibaraki Prefecture, Japan	Nuclear accident killed two workers and exposed one more to radiation levels above permissible limits
Saturday, February 16, 2002	Oak Harbor, Ohio, USA	Severe corrosion of control rod forces 24-month outage of Davis–Besse reactor
Monday, August 09, 2004	Fukui Prefecture, Japan	Steam explosion at Mihama Nuclear Power Plant killing four and injuring seven more
Tuesday, July 25, 2006	Forsmark, Sweden	An electrical fault at Forsmark Nuclear Power Plant caused one reactor to be shut down
Friday, March 11, 2011	Fukushima, Japan	Tsunami flooded and damaged five active reactor plants. Loss of backup electrical power led to overheating, meltdowns and evacuations

Source Sovacool (2008, 2010, 2011) and Sovacool et al. (2016)

LED may save 348 terawatt hours of electricity. In other words, this would mean saving USD 30 billion from not constructing new power plants (348 TW = annual power generation by 44 power plants of 1000 MW each) (Kunzig and Locatelli 2015)

Supercritical Coal-Based Power Plants

In an attempt to improve the energy efficiency and reduce CO_2 emission, countries like Germany, Japan and Denmark have introduced new technologies to extract more energy per unit of coal consumed through high heat and high pressure. As a result, the same amount of coal produces 30–40% more energy and a 4000 MW supercritical plant can save 4 million tons of coal each year. This technology, though expensive to begin with, will be useful for Asian countries like China and India that will continue to rely heavily on coal in for at least 30 years as Asia is projected to consume 80% of global coal consumption by 2040.

Metamaterials

The desire of man to change the world through conversion of matter into new, different and useful forms has led us to the golden age for materials. New alloys and composite materials with new ideas are emerging in the market. These 'smart' materials are energy efficient and high performers, e.g.

Perovskite: It is a crystal that can be used for enhancing the performance of LED bulbs.

Iron Pyrite: Nanoparticles from this material can help in recharging batteries much faster.

Graphene: It is also known as wonder material, is a two-dimensional carbon material that is extremely light, strong, transparent and works as a semiconductor. Not in commercial use as yet but it has great potential to revolutionize computer chips, efficient solar cells and water purification membrane.

Carbon Fibre Composites: It is made from a thin filament of carbon woven into a cloth and cut and pressed into any shape, and the fibres are bound together with a plastic resin, cured by heat and pressure. The molecular structure of carbon compounds produces strong chemical bonds much like those of diamonds. The resulting structure is stronger than steel (but 50% lighter) and longer lasting and does not corrode. Carbon fibre is already in wide use by aircraft industry, and all state-of-the-art aircraft such as Boeing 787, Airbus A 380 are made with carbon fibre. These aircraft are fuel efficient and low in emission.

Neutrino: Neutrinos are extremely important for future energy for three reasons: they are abundant, have feeble mass (the mass of a neutrino is of the order of 1×10^{-37} kg) and have no charge. They can, therefore, travel through human bodies, rocks, planets and stars without any interaction, thereby unravel scientific mysteries and help in medical science and communication. Neutrinos are the second most widely occurring particle after photons and are so abundant that nearly 100 trillion of them pass through us every second unnoticed. Since neutrinos can pass right through the Earth, they may revolutionize the communication system that currently uses satellites and cables (which also means transmission losses). Besides, neutrinos may also help in detecting minerals and oil deposits deep inside the Earth and ocean as well as detect geological defects.

Hybrid Cars

A hybrid car is eco-friendly, fuel efficient (30% more efficient) and less carbon (30% less carbon)-emitting vehicle that uses energy from petrol/diesel and electric motor.

5.6.1 Paradigm Shift Is Inevitable

The old paradigm of energy consumption and use was based on the adversarial harnessing belief that supply had to be met irrespective of environmental harmony, sustainability, equity and social justice. It was based on the assumption that economic growth was directly linked to energy consumption and ignored energy conservation, energy efficiency, environmental and health impacts.

Development = Economic Growth = Energy Consumption = Energy Demand Projection = Energy Supply Increase

It was the oil crisis of the 1970s, nuclear catastrophes in nuclear plants, the environmental degradation due to hydro and thermal plants and the global warming that compelled the nations to change the conventional energy paradigm dominating the thinking of all energy decision makers and planners to suggest a way out of the crisis through a new paradigm for energy.

The new paradigm recognized that what the individual and collective activities of human beings required was not energy per se but the work that the energy performs and the services that the energy provides, for example, illumination, warmth, cool, mobility. This approach gives precedence to **energy services** over the magnitude of energy consumption. While accepting that economic development does require a substantial increase in energy consumption and supply, the new paradigms realize that inefficiency in energy management has been the largest and expensive source of carbon emission and therefore suggest that improving **energy efficiency** of the devices that provide energy services and also the energy carriers such increases can be achieved in a sustainable way. This is possible through the innovation brought in by science and technology as well as delivery of new technology to the last consumer. The earlier this happens in any nation, the better that nation will be in terms of energy efficiency and the externalities associated with energy generation, transmission and end use. Historical data indicates that countries that brought in the Industrial Revolution during the nineteenth century (UK, USA Germany, France and Japan) showed a steep increase in energy intensities in the early phase that decreased progressively overtime. Several reasons can be attributed to this including improved raw material and finished goods, minimizing energy losses, reducing cost, increasing human skill, capacity and efficiency.

There has been a considerable shift in the energy production and technology of the developed world after the oil crisis of the 1970s, and a number of old energy-inefficient technologies have been replaced. This has not happened in the developing and underdeveloped world, as they are heavily dependent on technology import from the industrialized countries that invariably pass on the obsolete technology to the poor nations. But with the fast-changing current global scenario, and the concern for global warming impacts, there is a need to establish a comprehensive strategy for energy demand, production and efficiency.

Projects that improve the efficiency of energy both on the supply side and the demand side will have tremendous developmental benefits and future energy savings. Supply-side energy efficiency includes efficacy in burning of oil, gas and coal as well as production, transmission and distribution of electricity. For example, efficiency in conventional coal-based power plants can prevent 35% losses due to wastage in transmission, distribution and voltage adjustment. China has increased the energy efficiency of coal-fired thermal plants by 15% by introducing supercritical and ultra-supercritical technologies. Energy efficiency measures adopted by the USA in the 1970s saved around $365 billion in 30 years' time with the current conversion of primary energy into useful services being as high as 37–28% in many countries. Similarly, efficiency on the demand side is equally important especially in improving the transport, construction, housing (that accounts for 40% of energy consumption) and manufacturing sector that form the backbone of the global economy. Most of the demand for illumination, heating and cooling for human comforts, be it in cars, buses, houses, offices, factories, etc., can be optimized through better design, space management, judicious placement of doors and windows, passive solar heating, insulation, ventilation, equipment and time management. For instance, large buildings like shopping malls, airports, offices can make use of sunlight by simple modifications in design. Such buildings should mandatorily follow '**zero energy, zero carbon' code** meaning thereby that they will produce their own energy on site, preferably through renewable sources and emit no CO_2. Besides, simple actions like trees planting around such buildings and along highways and motorways can bring down temperature by a few degrees that can substantially cut down cooling requirement of houses and vehicles.

According to International Energy Agency, the investment in energy efficiency is rapidly growing, and in 2011, companies and governments in 11 countries had invested $11 billion which is almost equivalent to investment in electricity generation from coal, gas and oil. As a result, the revenue of companies in the USA grew by 20% a year ($7 billion) in the decade of 2011. Similarly, in China the revenue grew from zero in 2005 to $12 billion in 2010.

Besides efficiency, it is equally important to reduce the energy-related carbon dioxide emission. Carbon emissions from energy are based on total energy consumption and its **carbon intensity,** i.e. the units of CO_2 produced by a unit of energy consumed. The carbon intensity varies depending on the source of energy generation such as coal, gas and petroleum as well as the carbon policy. A low carbon use policy that reduces **energy intensity** (energy consumed per dollar of gross domestic product) by increasing energy efficiency and switching to low carbon lifestyle can go a long way in reducing emissions. An extremely difficult proposition considering the fact that global economy is poised to quadruple by 2050 and the current emission trends suggest a potentially catastrophic trajectory for carbon dioxide leading to 5 °C increase in temperature (compared to preindustrial period). Developed countries are to be discredited for this as they consume more than five times energy per capita vis-a-vis developing countries and release nearly two-third energy-related CO_2. If the global warming has to stay close to 2 °C warming levels, the developed countries will have to increase efficiency lest they prevent the economic growth of developing and low-income countries.

Despite technological advancement and better understanding of energy management, the current consumption pattern and comfort zone living by the urban population lead to an **'energy–temperature trap'** whereby rising temperatures increase the demand for cooling and increased cooling lead to rising temperatures. For example, the use of air conditioning by millions of cars on roads generates so much heat in the atmosphere that people are forced to use air conditions at home. At the same time, use of billions of air conditioners at home generates so much heat that car travellers are forced to switch on their air conditioners.

Arresting this process of CO_2 and temperature rise, the global community needs to:

a. Shift the energy mix from fossil fuel to renewable sources;
b. Substantial cost reduction in renewable energy technologies thereby ensuring widespread use;
c. Increasing the cost of non-renewable energy;
d. Motivating/compelling urban population to adopt energy mix with a greater share of renewable especially solar water heaters, solar cookers, solar lamps and so on;
e. Encourage/compel commuters to use public transport system;
f. Subsidize low/zero carbon emission fuels.

An insurmountable task ahead as urban and peri-urban centres will increase in number and expanse adding nearly 2 billion people in next 20 years. More and more new buildings will pop up, and 2.3 billion cars and an unknown number of heavy vehicles will be added. The world is going to face an intricate situation. The developed countries, on average, consume five times more energy per capita than developing countries. However, in order to improve the economy, the developing and less developed countries will have to use more energy, possibly through the cheaper and easier options of using sources such as coal and hydro.

5.6.2 Bottlenecks to Sustainability

One good thing that has happened in the twenty-first century is the acceptance of the fact that the conventional approach to energy planning was irrational, as it neither took care of the real ecological impact nor the distribution of benefits, nor the efficiency of services/output per unit of energy input. This often resulted in the skewed distribution of benefits especially to the poor and downtrodden. The conventional planning is also responsible for the current ecological development trap, as the energy that is being produced is having serious side effects including environmental degradation through heavy and uncontrolled siltation (hydel projects), loss of biodiversity, mining and pollution (thermal projects), and diversion of natural forests. The people who are located at the site of these development projects are invariably displaced and remain miserable thereafter. The other important factors are delay and cost overrun in case of power projects making it more expensive to produce the extra unit of energy.

The emphasis, now, needs a shift from the consumption of energy to the provision of energy services, and the true indicator of the sustainable energy development should be the level of energy services enjoyed by the population, particularly the rural poor, along with the magnitude of per capita energy consumption. There should be a paradigm shift from 'sustainable growth and development' to 'sustainable energy development and energy services'. This may be easier said than implemented on the ground. The fact remains that the cost of renewable such as wind turbines and solar panels is still prohibitive, and the technology transfer, as well as national capacity enhancement, is an expensive and time-consuming process. There are definite chances that many, if not all, countries will continue to use conventional sources irrespective of the climate consequences. **It is a trap, and most people are willing to fall into it rather than paying high cost for renewable energy**. Even the rich countries are seized with the dilemma of opting between green energy policies and conventional one. It is hard to sell them non-conventional energy with higher cost especially when the jobs are being cut, wages reduced and inflation escalating, and the rest of the world is doing little to cut their emissions. Experts say that, by 2030, Europe must derive 45% energy from renewable (with 55% emission cut) in order to meet the goal of cutting GHG emission by 80–95% in 2050. This seems unlikely with the current policy of 20-20-20 which means that by the year 2020 the EU members should reduce GHG emissions by 20% with 20% of the mix produced from renewable sources and a 20% improvement in energy efficiency. The 20-20-20 policy is a clear indication that achieving the target of 45% from renewable is too ambitious to be achieved by 2030. And more so when countries like France are committed to nuclear energy and bans shale gas exploration, Britain goes all out for shale gas and nuclear and Germany continues to burn coal with a promise to give up nuclear power.

While Japan switched off its last nuclear reactor in September 2013, it requires an additional $93 billion to meet the cost of importing extra oil, gas and coal that further threaten its ongoing trade deficit. The issue of safe energy is being debated with arguments for and against nuclear energy. Supporters of nuclear energy argue that per unit area production of electricity is very high in case of nuclear power plant and that switching to oil, gas and coal will further aggravate the GHG emission. It is also argued that installing solar panels and wind turbines in the mountainous areas will be an impossible task.

China, the emerging global economic power, produces 758 GW (out of total production of 1145 GW in 2012) from coal-based thermal plants. Despite well-established environmental consequences, China considers hydropower as green energy and has constructed thousands of large dams since 1949 to meet its energy needs. USA, the largest producer of oil, until the early 1970s when Arab states imposed oil embargo on Israel's Western allies in the aftermath of the 1967 war. In order to conserve its own resources, the US oil production went down sharply from 9.6 million barrels a day in 1970 to 5 million barrels a day in 2008. It imported huge amount crude oil from Canada, Saudi Arabia, Iraq, Libya and other countries to run its own economy and conserve its own oil. In the meanwhile, new technologies of hydraulic fracturing and horizontal drillings were evolved to tap shale gas and oil. USA has now increased its oil production to 7.4 million barrels a day and expects to return to 1970 level by 2019. This has been done to improve its economy by creating more jobs, reducing oil imports and increasing savings.

Foregoing examples of developed nations' unrestricted use of conventional energy sources are a clear indicator that the cost of energy from renewable sources such as wind and solar is still high, and therefore, market forces are not willing to switch over. Under the circumstance, the developing economies and, in particular, India and China are thirsty for oil and are expected to consume nearly 34 million barrels per day by 2025. With nearly a third of the global population to be fed, developed and economically secured, these two nations will continue to use oil and coal irrespective of the perceived ruins. Despite having great potential, the present grid-connected solar and installed wind power capacity in India is a meagre 2500 and 20,000 megawatts, respectively. With the demand for electricity alone expected to cross 950 GW by 2030, India will necessarily rely on conventional energy for a long time to maintain a growth rate between 10 and 12% in order to qualify for developed country status.

5.6.3 Back to Basics

Howsoever inane it may appear, but the truth remains that in many poor countries of Asia and Africa, animal power is still a major source of energy. Animals are also used on a limited scale in many advanced countries for tourism and adventure (sledges). Energy from animals should not be considered as a sign of backwardness. It is a wonderful, time-tested and most reliable renewable energy source capable of reducing pollution, carbon emission and catastrophic threats. If countries like the Netherlands can encourage cycles as an alternative to cars and buses, why cannot Asian and African countries encourage draught power. More than 84 million draught animals are being used in the agriculture sector in India contributing an energy equivalent of 30,000 MW annually or nearly half the installed capacity in the country. They plough 66 per cent of the country's farmland and are economical for small fragmented land (less than 4 ha). To replace them, India would require 15 million tractors, 6 million tons of fuel (diesel/petrol) and truck or trolleys worth Rs 300,000 million. A resource-starved country like India can ill afford energy switchover at this stage. What is required is investment in improving genetic breed of draught animals, designs of the cart and plough that are efficient to use, cost effective and easy to maintain. Special healthcare units and fodder banks would be required for the health and maintenance of these animals in the villages, as is done for any industry. All countries in the world especially the ones receiving good sunshine must make use of 'solar cooker' as excellent and effortless device to cook food. Dubbed as 'solar box cooker' by two American housewives in 1976, the present 'solar cooker' has paved the way for Third World kitchen revolution. It is a simple, practical, cheap, easily manageable cooking bonanza that retains the food value and ensures zero pollution. A box cooking for eight people saves two kilograms of charcoal and 6–7 kg of fuel-wood. It has found great acceptability in many Third World countries and saved many trees and coppice shoots from being slaughtered.

Rapid advancement towards urbanization and smart city development calls for 'zero energy import approach' where these cities generate captive energy by using all mechanisms and technologies that produce renewable energy and zero carbon such as solar water heaters in every building, solar lights, solar cooker, energy from human, animal and plant waste. The world's richest and most powerful nation like the USA also generates electricity from annual waste which is nearly 250 million tons. As a part of waste treatment strategy, this waste is compressed and used at modern landfill sites. As the organic waste breaks down, it releases carbon dioxide and methane which is then used to generate electricity. More than 600 energy projects are currently in operation that generates 15 billion kWh of electricity for sufficient to sustain one million homes a year.

The world, both rich and poor, is passing through a difficult transitory phase where an attempt is being made to bring in new technologies for safe energy generation. This may take between 25 and 50 years or may be more. This intervening phase will be crucial for all of us including conservationists, economists and scientists to ensure that the new world energy order is environmentally and economically secure.

References

Berckmans, Gert, Maarten Messagie, Jelle Smekens, Noshin Omar, Lieselot Vanhaverbeke, and Joeri Van Mierlo. 2017. Cost projection of state of the art lithium-ion batteries for electric vehicles up to 2030. *Energies* 10 (9): 1314.

Borthomieu, Yannick. 2014. Satellite lithium-ion batteries. In *Lithium-ion batteries*, 311–344.

Brekke, Kjell Arne, and Olof Johansson-Stenman. 2008. The behavioural economics of climate change. *Oxford Review of Economic Policy* 24 (2): 280–297.

Earth Policy Institute. 2008a Threatened species in major groups of organisms, 2007. Compiled by earth policy institute with 1980–1994 data from Worldwatch Institute, Signposts 2004, CD-ROM (Washington, DC: 2004); 1995 data from Global Wind Energy Council (GWEC), Global Wind 2006 Report (Brussels: 2007); 1996–2007 data from GWEC, "U.S., China, & Spain Lead World Wind Power Market in 2007," press release (Brussels: 6 February 2008).

Earth Policy Institute. 2008b. World cumulative installed geothermal electricity-generating capacity, 1950–2005. Compiled by earth policy institute with 1950–1999 data from Worldwatch Institute, Signposts 2004, CD-ROM (Washington, DC: 2004); 2000–2005 data from Eric Martinot, Tsinghua-BP Clean Energy Research and Education Center, Earth Policy Institute (12 April 2007).

Frei, Christoph, Rob Whitney, Hans-Wilhelm Schiffer, Karl Rose, Dan A. Rieser, Ayed Al-Qahtani, Philip Thomas et al. 2013. *World energy scenarios: Composing energy futures to 2050*. No. INIS-FR–14-0059. Conseil Francais de l'energie.

Gerbens-Leenes, P. W., A. R. Van Lienden, Arjen Ysbert Hoekstra, and Th H. Van der Meer. 2012. Biofuel scenarios in a water perspective: The global blue and green water footprint of road transport in 2030. *Global Environmental Change* 22 (3): 764–775.

Green, Richard, and Nicholas Vasilakos. 2011. The economics of offshore wind. *Energy Policy* 39 (2): 496–502.

International Energy Agency (IEA). 2013. *Key world energy statistics*. International Energy Agency.

Johnson, George. 2009. Plugging into the sun. *National Geographic* 216 (3): 28–53.

Kalam, APJ Abdul, and Srijan Pal Singh. 2011 Nuclear power is our gateway to a prosperous future. *The Hindu* 6: 10–11.

Kunzig, Robert, and L. Locatelli. 2015. Germany could be a model for how we'll get power in the future. *National Geographic* 15.

Li, Li, Yalin Lei, Chunyan He, Sanmang Wu, and Jiabin Chen. 2016. Prediction on the peak of the CO_2 emissions in China using the STIRPAT model. *Advances in Meteorology* 2016.

Mason, Colin. 2013. *The 2030 spike: Countdown to global catastrophe*, Routledge.

Meinshausen, Malte, Nicolai Meinshausen, William Hare, Sarah CB Raper, Katja Frieler, Reto Knutti, David J. Frame, and Myles R. Allen. 2009. Greenhouse-gas emission targets for limiting global warming to 2 C. *Nature* 458 (7242): 1158.

Mikolajczak, Celina, Michael Kahn, Kevin White, and Richard Thomas Long. 2012. *Lithium-ion batteries hazard and use assessment*. Springer Science & Business Media.

Parfomak, Paul W., Peter Folger, and Adam Vann. 2007. *Carbon dioxide (CO_2) pipelines for carbon sequestration: Emerging policy issues*. Washington, DC: Congressional Research Service, Library of Congress.

Petroleum, British. 2007. BP statistical review of world energy.

Powell, Joshua D. 2015. From pandemic preparedness to biofuel production: Tobacco finds its biotechnology niche in North America. *Agriculture* 5 (4): 901–917.

Ren21, Renewables. 2016 Global status report. *Renewable energy policy network for the 21st century*. http://www.ren21.net. Accessed 19 (2016).

Singh, Ajay Kumar, and Nalnish Chandr Singha. 2016. Environmental impact of nuclear power: Law and policy measures in India. *Humanities & Social Sciences Reviews* 4 (2): 88–95.

Sovacool, Benjamin K. 2008. The costs of failure: A preliminary assessment of major energy accidents, 1907–2007. *Energy Policy* 36 (5): 1802–1820.

Sovacool, Benjamin K. 2010. A critical evaluation of nuclear power and renewable electricity in Asia. *Journal of Contemporary Asia* 40 (3): 369–400.

Sovacool, Benjamin K. 2011. *Contesting the future of nuclear power: A critical global assessment of atomic energy*. World Scientific.

Sovacool, Benjamin K., Rasmus Andersen, Steven Sorensen, Kenneth Sorensen, Victor Tienda, Arturas Vainorius, Oliver Marc Schirach, and Frans Bjørn-Thygesen. 2016. Balancing safety with sustainability: assessing the risk of accidents for modern low-carbon energy systems. *Journal of cleaner production* 112: 3952–3965.

Taylor, Lenore, and Tania Branigan. 2014. US and China strike deal on carbon cuts in push for global climate change pact. *The Guardian* 12.

The Economist. 2015a Charge of the lithium brigade. *The Economist*, 28 May 2015.

The Economist. 2015b. Half death; the future of nuclear energy. *The Economist*, 31 Oct 2015.

The Economist. 2015c. Why the Dutch oppose windmillsDutch Quixote. *The Economist*, 2nd July 2015| Amsterdam.

The New Climate Economy. 2014. 2011 Better growth, better climate. The new climate economy report, the global commission on the economy and climate.

World Bank. 2013. Development Economics Dept. Development Data Group, and World Bank. Environment Dept. *The little green data book...: From the world development indicators...* International Bank for Reconstruction and Development/The World Bank.

Printed in the United States
By Bookmasters